Maria Spichkova

Specification and Seamless Verification of Embedded Real-Time Systems

FOCUS on Isabelle

VDM Verlag Dr. Müller

Imprint

Bibliographic information by the German National Library: The German National Library lists this publication at the German National Bibliography; detailed bibliographic information is available on the Internet at http://dnb.d-nb.de.

Cover image: www.purestockx.com

Publisher:
VDM Verlag Dr. Müller Aktiengesellschaft & Co. KG, Dudweiler Landstr. 125 a, 66123 Saarbrücken, Germany,
Phone +49 681 9100-698, Fax +49 681 9100-988,
Email: info@vdm-verlag.de

Produced in USA and UK by:
Lightning Source Inc., La Vergne, Tennessee, USA
Lightning Source UK Ltd., Milton Keynes, UK
BookSurge LLC, 5341 Dorchester Road, Suite 16, North Charleston, SC 29418, USA

ISBN: 978-3-8364-9452-6

Maria Spichkova

Specification and Seamless Verification of Embedded Real-Time Systems

Abstract

The purpose of this thesis is to create a coupling of the formal specification framework FOCUS in the generic theorem prover Isabelle/HOL, a logical framework based on Higher-Order Logic. The main focus of this work is on specification and verification of systems that are especially safety critical – embedded real-time systems.

Isabelle/HOL is an interactive semi-automatic theorem prover and in the proofs of some system properties human time must be invested. These proofs strongly depend on the specifications of the system and the properties. By considering the framework "FOCUS on Isabelle", which is result of the coupling, we can influence the complexity of proofs already during the specification phase, e.g. reformulating specification to simplify the Isabelle/HOL proofs for a translated FOCUS specification. Thus, the specification and verification/validation methodologies are treated as a single, joined, methodology with the main focus on the specification part. This methodology uses particularly the idea of refinement-based verification, where the verification of system properties can be treated as a validation of a system specification with respect to the specification representing the properties.

The key contributions of the thesis are

✓ Deep embedding of that part of the framework FOCUS, which is appropriate for the specification of real-time systems, into Isabelle/HOL. "FOCUS on Isabelle" enables to validate and verify system specifications in a methodological way.

✓ Syntax extensions for FOCUS for the argumentation over time intervals: a special kind of tables, timed state transition diagrams, and a number of new operators. The deep embedding into Isabelle/HOL includes all these extensions.

✓ Schemata for automatic correctness proofs in Isabelle/HOL of the syntactic interfaces for specified system components.

The feasibility of the proposed approach was evaluated on three case studies that cover different application areas:

✓ Steam Boiler System (process control),

✓ FlexRay communication protocol (data transmission),

✓ Automotive-Gateway System (memory and processing components, data transmission).

The results of "FOCUS on Isabelle" can also be extended to a complementary approach, "JANUS on Isabelle", that presents a coupling of the formal specification framework for services, JANUS, with Isabelle/HOL.

Acknowledgements

I would like to thank all those people that have helped me, either directly or indirectly, with the development of this thesis.

My special thanks go to my supervisor Manfred Broy. He gave me the opportunity to work in a highly qualifed research group and guided my activity in the past four years. His own work provided much of the motivation of this thesis. Furthermore, I want to thank Manfred Broy for all his well-founded and helpful advice and support, as well as for directing me towards the concrete topic of my thesis.

Special thanks go also to Tobias Nipkow for being my referee and improving the quality of the realization with his technical advice.

Thanks are due to all my colleagues for years of discussion. In particular I am grateful to thank my colleagues Stefan Berghofer and Clemens Ballarin for carefully reading draft version of this thesis and suggesting numerous improvements. I also would like to thank Oscar Slotosch, Bernhard Rumpe, Alexander Ziegler, Bernhard Schätz, Daniel Ratiu, Christian Kühnel, Martin Wildmoser, and Borislav Gajanovic for inspiring and clarifying discussions.

Last but not least, I want to thank my mother for her continuous support and encouragement during the work.

Contents

1. Introduction   1
   1.1. Motivation   1
   1.2. Isabelle/HOL   3
   1.3. FOCUS   5
        1.3.1. Concept of Streams   6
        1.3.2. Operators on Streams   7
        1.3.3. Specification Styles 10
        1.3.4. Elementary and Composite Specifications 10
        1.3.5. Sheaves and Replications 13
        1.3.6. Refinement 14
        1.3.7. Causality 17
   1.4. JANUS 18
   1.5. Outline 18

2. FOCUS on Isabelle   21
   2.1. Specification of Embedded Real-Time Systems 22
   2.2. Stream Representation 23
   2.3. Representation of Datatypes 26
   2.4. FOCUS Operators Representation 28
        2.4.1.   nth Message of a Stream 28
        2.4.2.   Length of a Stream 29
        2.4.3.   Concatenation Operator 29
        2.4.4.   Prefix of a Stream 30
        2.4.5.   Truncate a Stream 31
        2.4.6.   Domain and Range of a Stream 32
        2.4.7.   Time-synchronous Stream 33
        2.4.8.   Make Untimed 34
        2.4.9.   Time Stamp Operator 34
        2.4.10. Filtering Operator 35
        2.4.11. Application Operator 36
        2.4.12. Stuttering Removal Operator 36
   2.5. FOCUS Operators: Extensions 37
        2.5.1.   Number of Time Intervals in a Finite Stream 37
        2.5.2.   Timed Truncation Operator 37
        2.5.3.   Time Interval 38
        2.5.4.   Drop a Stream 39
        2.5.5.   Timed Merge 40

2.5.6. Concatenation of Time Intervals 41
2.5.7. Limited Number of Messages 42
2.5.8. Stuttering Removal Operator for Timed Streams 44
2.5.9. Changing Time Granularity 44
2.5.10. Deleting the First Time Interval 46
2.5.11. First Nonempty Time Interval 46
2.5.12. Index of the First Nonempty Time Interval 48
2.5.13. Last Nonempty Time Interval 49
2.6. tiTable 50
2.7. Encapsulated States 54
2.8. Timed State Transition Diagrams 56
2.9. Mutually Recursive Functions 59
2.10. Time-Synchronous Streams 59
2.11. Isabelle/HOL Semantics of Elementary and Composite FOCUS
       Specifications 61
       2.11.1. Auxiliary Datatypes, Functions and Predicates 61
       2.11.2. Elementary Specification 62
       2.11.3. Composite Specification 65
       2.11.4. Relations between Sets of Channels 67
       2.11.5. Example: Steam Boiler 69
2.12. Composition types 73
       2.12.1. Sequential composition 73
       2.12.2. Parallel composition 74
       2.12.3. "Mix"-composition 75
       2.12.4. Loop 76
2.13. Sheaves and Replications 79
       2.13.1. Sheaves of Channels 79
       2.13.2. Specification Replications 82
       2.13.3. Isabelle/HOL Specification of Relations between Sets of
               Channels 84
       2.13.4. Disjoint Channels in a Sheaf 90
       2.13.5. Example: System with Sheaves of Channels 91
2.14. Translation schema: From FOCUS to Isabelle/HOL 95
2.15. Summary 103

3. Specification and Verification Methodology   105
3.1. Refinement 105
     3.1.1. Refinement Layers of a Specification Group 106
     3.1.2. Behavioral Refinement 106
     3.1.3. Interface Refinement 109
     3.1.4. Conditional Refinement 109
3.2. Consistency of a Specification 110
3.3. Refinement-Based Verification 111
3.4. Key Ideas of the Specification and Verification Methodology . . . 114
3.5. Specification Hints and Mistakes 118
3.6. Complementary Approach: JANUS on Isabelle 120

3.7. Summary 121

4. Case Studies 123

4.1. Steam Boiler System 123
    4.1.1. Datatypes 124
    4.1.2. Requirement Specification 124
    4.1.3. Architecture Specification 126
    4.1.4. Steam Boiler Component 127
    4.1.5. Converter Component 129
    4.1.6. Controller Component 130
    4.1.7. Verification of the Steam Boiler System 134
    4.1.8. Results of the Case Study 136
4.2. FlexRay Communication Protocol 137
    4.2.1. Datatypes 138
    4.2.2. Input/Output Relations between Channels 140
    4.2.3. Auxiliary Predicates 144
    4.2.4. Requirement Specification 148
    4.2.5. Architecture Specification 148
    4.2.6. Cable Component 150
    4.2.7. Controller Component 151
    4.2.8. Verification of the FlexRay System wrt. its Requirements 153
    4.2.9. Results of the Case Study 161
4.3. Automotive-Gateway 162
    4.3.1. Datatypes 163
    4.3.2. Input/Output Relations between Channels 164
    4.3.3. Gateway System: Architecture and Requirements 168
    4.3.4. ECall Service Center 170
    4.3.5. Gateway: Requirements Specification 171
    4.3.6. Gateway: Architecture Specification 172
    4.3.7. Sample Component 173
    4.3.8. Delay Component 181
    4.3.9. Loss Component 181
    4.3.10. Verification of the Gateway 182
    4.3.11. Verification of the Gateway System 184
    4.3.12. Extended Requirements Specification of the Gateway . . . 188
    4.3.13. Results of the Case Study 190
4.4. Summary 190

5. Conclusions 193

5.1. Summary 193
5.2. Outlook 195

Bibliography 197

A. Isabelle Definitions and Lemmas about FOCUS Operators   201

A.1. Theory stream.thy (FOCUS streams) 201
     A.1.1. Lemmas about inf_truncate 213
     A.1.2. Lemmas about fin_make_untimed 213
     A.1.3. Lemmas about inf_disj and inf_disjS 215
A.2. Theory join_ti – Concatenation of time intervals 215
A.3. Changing Time Granularity 216
A.4. Theory ArithExtras.thy 221
A.5. Theory ListExtras.thy 221
A.6. Auxiliary Arithmetic Lemmas 223

B. Isabelle/HOL Specifications and Proofs   226

B.1. Steam Boiler System Specification 226
B.2. Proof of the Steam Boiler System Properties 227
     B.2.1. Properties of Controller Component 227
     B.2.2. Properties of the System 228
     B.2.3. Proof of the Refinement Relation 230
B.3. Theory FR - System Specification 230
     B.3.1. Auxiliary predicates 230
     B.3.2. Main definitions 231
B.4. Proof of the FlexRay System Properties 233
B.5. Automotive-Gateway System Specification 238
B.6. Auxiliary Proofs for the Automotive-Gateway System 244
     B.6.1. Properties of the defined datatypes 244
     B.6.2. Equivalence of the titable representations: SampleT and
         SampleT_ext 245
     B.6.3. Auxiliary Lemmas 245
B.7. Proof of the Automotive-Gateway System Properties 255
     B.7.1. Properties of the Gateway 256
     B.7.2. Proof of the Refinement Relation for the Gateway Requirements 262
     B.7.3. Lemmas about Gateway Requirements 263
     B.7.4. Properties of the Gateway System 264
     B.7.5. Proof of the Refinement for the Gateway System 268
     B.7.6. Proof of the Refinement Relation for the Extended Gateway Requirements 269
     B.7.7. Lemma about Extended Gateway Requirements 269
     B.7.8. Proof of the Refinement for the Gateway System (Based on the Extended Gateway Requirements) 272

List of Figures

1.1. FOCUS Specification Frames and Styles: Elementary Specifications 11
1.2. FOCUS Specification Styles: Composite Specifications 12
1.3. Interface Refinement 16
1.4. Conditional Interface Refinement 17

2.1. Duplicated time raster 45
2.2. Wrong and correct specification of subcomponents 67
2.3. Architecture of the Steam Boiler System 70

3.1. Refinement Layers of a Specification Group S 107
3.2. Adding new requirement R to the list L of requirements of the specification S, $L \cup \{R\}$ 114
3.3. Specification Group S: Representation of the whole group by a joined Isabelle/HOL theory 115
3.4. Specification Group S: Representation of every component by a *single* (separate) Isabelle/HOL theory 116
3.5. Specification Group S: Representation of all components of the same abstraction layer by a *joined* Isabelle/HOL theory 117

4.1. Timed state transition diagram for the component Sample 179

List of Figures

1. Introduction

1.1. Motivation

Embedded systems is one of the most challenging fields of systems engineering: such a system must meet real-time requirements, is safety critical and distributed. The current practice in industry of ensuring that a software system fulfills its requirements is testing. However, testing can only demonstrate the presence, but not the absence of errors. Therefore, developing complex, safety critical systems it is insufficient only to test them. The increasing combinatorial complexity and the safety and quality requirements of embedded real-time systems implies that we need another solution to the problem. It is now widely recognized that only formal methods can provide the level of assurance required by the increasing complexity and the high safety and quality requirements to such systems, because these allow not only to test correctness and safety, which is not enough for such kinds of interactive systems, but also to prove that the requirements are met: verification guarantees fulfillment of the requirements. Coupling a specification framework with some verification system will reduce the lavishness and error-proneness of system specifications.

For the development of embedded real-time systems in most cases experts of different disciplines have to cooperate, and for such a cooperation a specification of the developing system, i.e. precise and detailed description of its behavior and/or structure, is important. Embedded system are real, but their behaviors are mathematical objects and about mathematical objects one can argue formally. One aim of formal methods is to prove or to automatically evaluate behavior properties of a system in a systematic way, based on a clear mathematical theory. A formal specification is in general more precise than a natural language one, but it can also contain mistakes or disagree with requirements. Therefore, for safety critical systems it is not enough to have detached formal specifications – for this case verification is needed. This is the only way to make sure that the specification conforms to its requirements and is consistent.

This approach introduces a coupling of a specification framework with a verification system. Given a system, represented in a formal specification framework, one can verify its properties by translating the specification to a Higher-Order Logic and subsequently using the theorem prover Isabelle/HOL (or the point of disagreement will be found). Moreover, using this approach one can validate the refinement relation between two given systems. The approach uses particularly the idea of refinement-based verification, where a verification of system properties can be treated as a validation of a system specification with respect to the specification of the properties.

To design a large (real) system a number of refinement steps are needed – it is in most cases even impossible to make a concrete specification (implementation) from its abstract requirements specification in a single step. A system development process has several levels of abstraction. In most cases a more abstract specification is more readable and easier to understand, as well as it has more chances to be reused. Thus, designing a system, a refinement relation must be shown not only between requirements and architecture specifications, but for every step on which a more abstract specification is refined to a more precise one. The proofs of refinement relations between specifications of neighbor levels of abstractions are in general simpler and shorter than the proof of the refinement relation between the most abstract and the most concrete specifications.

In order to design systems in a step-wise, modular style we use Focus [BS01], a framework for formal specifications and development of interactive systems. This framework provides a number of specification techniques for distributed systems and concepts of refinement. Formal specifications of real-life systems can become very large and complex, and are as a result hard to read and to understand. Therefore, it is too complicated to start the specification process in some low-level framework, First-Order or Higher-Order Logic etc. directly. To avoid this problem Focus supports a graphical specification style based on tables and diagrams. Focus is preferred here over other specification frameworks also since it has an integrated notion of time and modeling techniques for unbounded networks (specification replications, sheaves of channels). For example, the B-method [Abr96] is used in many publications on fault-tolerant systems, but it has neither graphical representations nor integrated notion of time. Moreover, the B-method also is slightly more low-level and more focused on the refinement to code rather than formal specification.

The first attempt to represent the first version of the Focus syntax [BDD⁺92] (without representation of time, modeling techniques for unbounded networks, etc.) in a verification system was done by B. Schätz and K. Spies (see [SS95]). In this approach the HOLCF specialization of the theorem prover Isabelle was chosen. HOLCF (see [Reg94], [Reg95] and [MNvOS99]) is the definitional extension of Church's Higher-Order Logic with Scott's Logic for Computable Functions that has been implemented in Isabelle. HOLCF supports the standard domain theory but also coinductive arguments about lazy datatypes. The main disadvantage of using HOLCF in practice is the difficulty of logic understanding in comparison to HOL.

A number of methods for the implementation of interactive systems in the first version of Focus were described also by O. Slotosch (see [Slo97]).

The first attempt of coupling of Focus with an automatic verification system was done by J. Schumann and M. Breitling [SB99]. As the verification system was chosen SETHEO [LSBB92], an automatic theorem prover for proving the unsatisfiability of formulas in First-Order Clause Logic. This case study of J. Schumann and M. Breitling has shown that such a coupling is in principle possible, but there are also a number of problems and open questions.

In our approach we chose a prover for Higher-Order Logic, because the power of First-Order Logic is not enough to represent in a direct way several specifications of distributed interactive systems. As the verification system Isabelle/HOL we have chosen Isabelle/HOL [NPW02, Wen04], an interactive semi-automatic theorem prover for Higher-Order Logic. The disadvantage of only semi-automated proofs is compensated by the advantage of using Higher-Order Logic.

A mapping of operators in FOCUS to the corresponding definitions in HOL alone is not sufficient for the method to become easy. Because of this, we also present the specification and proof methodology – the main point in our methodology is an alignment on the future proofs to make them simpler and appropriate for application not only in theory but also in practice. For this we have performed a number of case studies, whose results have helped us to find out different problem points and corresponding solutions for the coupling FOCUS and Isabelle/HOL. The proofs of some system properties can take considerable (human) time since the Isabelle/HOL is not fully automated. But considering the framework "FOCUS on Isabelle", which is presented here, we can influence on the complexity of proofs already doing the specification of systems and their properties, e.g. modifying (reformulating) specification to simplify the Isabelle/HOL proofs for a translated FOCUS specification. Thus, the specification and verification/validation methodologies are treated as a single, joined, methodology with the main focus on the specification part.

The thesis presents not only the approach of coupling of FOCUS with Isabelle/HOL (as deep embedding), namely "FOCUS on Isabelle", but also a complementary approach, "JANUS on Isabelle", that presents a coupling of a JANUS with Isabelle/HOL. JANUS [Bro05] is a specification framework for services, that is build on the base of FOCUS and uses different, but similar syntax and semantics.

The remainder of this chapter gives an overview of the thesis: Sections 1.2, 1.3 and 1.4 introduce the main concepts of Isabelle/HOL, FOCUS and JANUS respectively, and Section 1.5 gives an overview over the remainder of the thesis.

1.2. Isabelle/HOL

Isabelle [NPW02] is a specification and verification system implemented in the functional programming language ML. Isabelle/HOL is the specialization of Isabelle for Higher-Order Logic. To specify a system with Isabelle is done by creating *theories*. Isabelle/HOL allows also to deal with induction without additional efforts. A theory is a named collection of types, functions (constants), and theorems (lemmas). Similar to the module concept from FOCUS, we can understand a theory in Isabelle as a module. The general format of a theory T is

```
theory T = Main + B1 + ...+ Bn:
    declarations, definitions, lemmas with proofs
end
```

where *B1*, ..., *Bn* are the names of existing theories that T is based on and *declarations, definitions, lemmas with proofs* represents the newly introduced concepts.

The base types in Isabelle/HOL are *bool*, the type of truth values and *nat*, the type of natural numbers. The base type constructors are *list*, the type of lists, and *set*, the type of sets. Function types are denoted by $\Rightarrow$. The operator $\Rightarrow$ is right-associative. The type variables are denoted by *'a*, *'b* etc.

A new datatype can be defined using keyword *datatype*. The general form of the definition is

$$datatype\ (\alpha_1, \ldots, \alpha_n)\ t = C_1\ \tau_{11} \ldots \tau_{1k_1}\ |\ \cdots\ |\ C_m\ \tau_{m1} \ldots \tau_{mk_m}$$

where α_i are distinct type variables, C_i are distinct constructor names and τij are (defined) types.

Type synonyms are created by a **types** command:

$$types\ number = nat$$

Terms in Isabelle/HOL are formed as in functional programming by applying functions to arguments. Terms may also contain λ-abstractions.

The notation $t :: \tau$ is used, if for a term t an explicit type constraint τ is needed.

Isabelle/HOL supports also some basic constructs from functional programming, like

$$if\ b\ then\ t_1\ else\ t_2$$

$$let\ x = t\ in\ u\ \text{(is equivalent to } u \text{ where all occurrences of } x$$
$$\text{are replaced by } t)$$

$$case\ e\ of\ c_1 \Rightarrow e_1\ |\ \cdots\ |\ c_n \Rightarrow e_n\ \text{(evaluates to } e_i \text{ if } e \text{ is of the form } c_i)$$

A recursive function or predicate can be declared in Isabelle/HOL by the keyword **consts**:

$$consts\ function\_name :: function\_type$$

The keyword **primrec** indicates that the recursion in the definition is of particularly primitive kind.

Arbitrary total recursive functions can be defined by means of **recdef**[1] (see [NPW02]): recursion need not involve datatypes, termination is proved by showing that the arguments of all recursive calls are smaller in a suitable sense.

Constants are defined in Isabelle/HOL by the keyword **constdefs**, e.g.

$$constdefs\ xor :: bool \Rightarrow bool \Rightarrow bool$$
$$xor\_def\ A\ B \equiv (A \wedge \neg B) \vee (\neg A \wedge B)$$

A theorem (lemma) can be proved, e.g. in the following ways.

---

[1] This kind of definitions is in general more complicated that the **primrec** definition: termination function must be defined explicitly, in most cases Isabelle/HOL needs also some (nontrivial) **hints** to prove the termination. If it is possible, it is better to restrict the definition of a Focus function or predicate to the primitive kind.

✓ Automatically:
apply auto
apply clarify
etc.

✓ Theorems about recursive functions are proved by induction (on the argument *i* on which the function is defined by recursion):
apply (induct i) (for *recdef*-definitions - *induct_tac* instead of *induct*)

✓ Using simplification rules, adding to the simplification rules function definitions, or ignoring the theorem assumptions (to avoid loops) by simplification, etc.:

◇ *apply simp*

◇ *apply (simp add: Let_def)* (expand *let*-construct)

◇ *apply (simp add: xor_def)* (add the definition of the *xor* function to the simplification)

◇ *apply (simp (no_asm))* (assumptions are not simplified and not used in the simplification of the conclusion)

◇ *apply (simp (no_asm_simp))* (assumptions are not simplified, but are used in the simplification of the conclusion)

◇ *apply (simp (no_asm_use))* (assumptions are simplified, but are not used in the simplification of each other or the conclusion)

✓ Using automatic case splits, like
apply (split split_if) (split *if*-expression)

✓ Using rules of natural deduction.

✓ Generating new subgoal (the corresponding new assumption will be added to the current goal)
apply (subgoal_tac "new assumption")

✓ Using predefined theorems or lemmas.

A large collection of predefined Isabelle/HOL theories, e.g. sets, ordered rings and fields, exponentiations, division operators etc, is presented on the Isabelle/HOL official web site http://isabelle.in.tum.de/library/index.html.

For detailed description of Isabelle/HOL see [NPW02] and [Wen04].

1.3. FOCUS

FOCUS [BS01] is a framework for formal specifications and development of distributed interactive systems. A distributed system in FOCUS is represented by its components that are connected by communication lines called *channels*, and are described in terms of its input/output behavior, which is total. The components can interact and also work independently of each other. A specification can be elementary or composite – composite specifications are built hierarchically from the elementary ones. Elementary specifications are divided into untimed, timed, and time-synchronous according their level of time abstraction (see Section 1.3.1).

The channels in this specification framework are *asynchronous communication links* without delays. They are *directed*, *reliable*, and *order preserving*. Via these channels components exchange information in terms of *messages* of specified types. Messages are passed along the channels one after the other and delivered in exactly the same order in which they were sent.

In FOCUS any specification characterizes the relation between the *communication histories* for the external *input* and *output channels*. The formal meaning of a specification is exactly this external *input/output relation*. The FOCUS specifications can be structured into a number of formulas each characterizing a different kind of property, the most prominent classes of them are *safety* and *liveness properties*.

1.3.1. Concept of Streams

The central concept in FOCUS are *streams*, that represent communication histories of *directed channels*. For any set of messages M, M^ω denotes the set of all streams, M^∞ and M^* denote the sets of all infinite and all finite streams over the set M respectively.

$$M^\omega \stackrel{\text{def}}{=} M^* \cup M^\infty$$
$$M^* \stackrel{\text{def}}{=} \bigcup_{n \in \mathbb{N}}([1..n] \to M)$$
$$M^\infty \stackrel{\text{def}}{=} \mathbb{N}_+ \to M$$

In FOCUS streams are represented as functions mapping natural numbers to messages, where a message can be for the case of untimed stream only a data message of some type and for the case of timed stream either a data message or *time tick* (represented by $\sqrt{}$).

An empty stream is denoted in FOCUS by $\langle \rangle$.

$M^{\underline{\omega}}$ denotes in FOCUS the set of all timed streams, $M^{\underline{\infty}}$ and $M^{\underline{*}}$ denote the sets of all infinite and all finite timed streams over the set M respectively:

$$M^{\underline{\omega}} = M^{\underline{*}} \cup M^{\underline{\infty}}$$
$$M^{\underline{*}} \stackrel{\text{def}}{=} \bigcup_{n \in \mathbb{N}}([1..n] \to M \cup \{\sqrt{}\})$$
$$M^{\underline{\infty}} \stackrel{\text{def}}{=} \mathbb{N}_+ \to M \cup \{\sqrt{}\}$$

Defined in this way, streams are functions mapping the indexes in their domains to their messages. A *timed stream* is represented by a sequence of messages and *time ticks*, the messages are also listed in their order of transmission. The ticks model a discrete notion of time (see also Section 1.4).

1.3.2. Operators on Streams

The *domain* and the *range* of a stream are defined in FOCUS like for the functions:

$$\text{dom} \in M^{\omega} \to \{[1, \ldots, n] \mid n \in \mathbb{N}\} \cup \{\mathbb{N}_{\infty}\}$$

$$\text{rng} \in M^{\omega} \to \mathbb{P}(M)$$

The *length operator* for streams

$$\# \in M^{\omega} \to \mathbb{N}_{\infty}$$

yields the length of the stream to which is applied:

$$\#s = k \Leftrightarrow s \in [1, \ldots, k] \to M$$

The nth message of a stream s can be denoted in FOCUS either by $s(n)$ or by $s.n$. We prefer the second kind of notation.

Another basic FOCUS operator on streams is the *concatenation operator*:

$$\frown \; \in \; M^{\omega} \times M^{\omega} \to M^{\omega}$$

The concatenation of two streams produces a stream that starts with the messages of the first stream followed by the messages of the second stream:

$$(s \frown r).k \;\stackrel{def}{=}\; \begin{cases} s.k & if \;\; 1 \le k \le \#s \\ r.(k - \#s) & if \;\; \#s < k \le \#s + \#r \end{cases}$$

By this definition it follows that

$$\#s = \infty \;\Rightarrow\; s \frown r = s$$

The append function $m \,\&\, s$ results concatenation of the message m to the head of stream s:

$$m \,\&\, s \;\stackrel{def}{=}\; \langle m \rangle \frown s$$

The *first* and *rest* operators yield the first element of the stream and the stream without the first element respectively:

$$\text{ft} \in M^{\omega} \to M \cup \{\bot\}, \qquad \text{rt} \in M^{\omega} \to M^{\omega}$$

$$\text{ft}.s \;\stackrel{def}{=}\; \begin{cases} \bot & if \; \#s = 0 \\ s.1 & \text{otherwise} \end{cases}$$

$$\text{rt}.s \;\stackrel{def}{=}\; \begin{cases} \langle\rangle & if \; \#s = 0 \\ r & \text{otherwise} \;, \text{ where } s = (\text{ft}.s) \,\&\, r \end{cases}$$

A stream r is a prefix of a stream s if r is an initial segment of s or if r is equal to s. The *prefix ordering relation* is defined as follows:

$$\sqsubseteq \; \in M^{\omega} \times M^{\omega} \to \mathbb{B}\text{ool}$$

$$r \sqsubseteq s \;\stackrel{def}{=}\; \exists\, t \in M^{\omega} : r \frown t = s$$

The *truncation operator*

$$| \in M^\omega \times \mathbb{N}_\infty \to M^\omega$$

is used to truncate a stream at a certain length:

$$s|_j = \begin{cases} r & \text{if } 0 \le j \le \#s, \text{where } \#r = j \wedge r \sqsubseteq s \\ s & \text{otherwise} \end{cases}$$

The *filtering operator*

$$\text{\textcircled{s}} \in \mathbb{P}(M) \times M^\omega \to M^\omega$$

filters away messages that do not belong to the filtering set. $M \text{\textcircled{s}} s$ denotes the substream of s obtained by removing all messages in s that does not belong to the set M. The FOCUS definition for a finite stream s:

$$\begin{aligned} M \text{\textcircled{s}} \langle \rangle &= \langle \rangle \\ m \in M \Rightarrow \quad M \text{\textcircled{s}} m \,\&\, s &= m \,\&\, M \text{\textcircled{s}} s \\ m \notin M \Rightarrow \quad M \text{\textcircled{s}} m \,\&\, s &= M \text{\textcircled{s}} s \end{aligned}$$

For an infinite stream one additional equation is used:

$$(\text{rng}.s \cap M) = \{\} \Rightarrow M \text{\textcircled{s}} s = \langle \rangle$$

The *stuttering removal operator*

$$\propto \in M^\omega \to M^\omega$$

removes consecutive repetitions of messages:

$$\begin{aligned} \propto.\langle \rangle &= \langle \rangle \\ \propto.\langle m \rangle &= \langle m \rangle \\ \propto.\langle m \rangle^\infty &= \langle m \rangle \\ \propto.(m_1 \,\&\, m_2 \,\&\, s) &= \begin{cases} \propto.(m_2 \,\&\, s) & \text{if } m_1 = m_2 \\ m_1 \,\&\, \propto.(m_2 \,\&\, s) & \text{otherwise} \end{cases} \end{aligned}$$

The application operator

$$\text{map} \in M^\omega \times (M \to T) \to T^\omega$$

applies a function $g \in M \to T$ to each element of a stream:

$$\begin{aligned} \text{map}(\langle \rangle, g) &= \langle \rangle \\ \text{map}(m \,\&\, s, g) &= g(m) \,\&\, \text{map}(s, g) \end{aligned}$$

Time abstraction operator has the signature

$$\_ \in M^\omega \to M^\omega$$

and is formally defined by

$$\overline{s} = M \circledS s$$

$\overline{s}$ denotes in FOCUS the untimed stream obtained by removing all ticks in s. For example,

$$\overline{\langle m_1, \sqrt{}, m_2, m_3, \sqrt{}, m_4, m_5, m_6, \sqrt{}, m_7 \rangle} = \langle m_1, m_2, m_3, m_4, m_5, m_6, m_7 \rangle$$

The *timed truncation operator*

$$\downarrow \in M^{\infty} \times \mathbb{N}_{\infty} \to M^{\underline{*}}$$

truncates a timed *infinite* stream at a certain point in time:

$$s \downarrow_n \stackrel{\text{def}}{=} \begin{cases} s & \text{if } n = \infty \\ \langle \rangle & \text{if } n = 0 \\ r & \text{otherwise}, \text{ where } r \sqsubseteq s \wedge \#\{\sqrt{}\} \circledS r = n \wedge r.\#r = \sqrt{} \end{cases}$$

For example, let

$$s = \langle m_1, \sqrt{}, m_2, m_3, \sqrt{}, m_4, m_5, m_6, \sqrt{}, m_7, \sqrt{} \dots \rangle,$$

then

$$s \downarrow_0 = \langle \rangle$$
$$s \downarrow_1 = \langle m_1, \sqrt{} \rangle$$
$$s \downarrow_2 = \langle m_1, \sqrt{}, m_2, m_3, \sqrt{} \rangle$$
$$s \downarrow_3 = \langle m_1, \sqrt{}, m_2, m_3, \sqrt{}, m_4, m_5, m_6, \sqrt{} \rangle$$

$\dots$

Thus, the last message in the truncated stream $s \downarrow_n$ (if it is nonempty) is always a tick $\sqrt{}$.

The predicate

$$\mathsf{ts} \in M^{\infty} \to \mathbb{Bool}$$

holds for a timed stream s, iff s is time-synchronous in the sense that exactly one message is transmitted in each time interval:

$$\mathsf{ts}(s) \stackrel{\text{def}}{=} \forall j \in \mathsf{dom}.s : (\mathsf{even}(j) \Leftrightarrow s.j = \sqrt{})$$

The FOCUS time stamp operator $tm\ s\ k$ yields the time interval in which the kth message in the time stream s is transmitted.

$$k \in \mathsf{dom}.\overline{s} \implies \mathsf{tm}(s, k) \stackrel{\text{def}}{=} \min\{j \in \mathbb{N} \mid \#\overline{s \downarrow_j} \geq k\}$$

For example, let

$$s = \langle m_1, m_2, \sqrt{}, \sqrt{}, m_3, m_4, m_5 \rangle,$$

then

$$\mathsf{tm}(s, 1) = 1$$
$$\mathsf{tm}(s, 2) = 1$$
$$\mathsf{tm}(s, 3) = 3$$

1.3.3. Specification Styles

FOCUS supports a variety of *specification styles* which describe system components by logical formulas or by diagrams and tables representing logical formulas (see Figure 1.1):

✓ *Relational Style* – specification is expressed directly as a relation between the input and output streams.

✓ *Equational Style* – the behavior is described as equations, the specifications look very much like programs coded in a functional programming language, this style is suited for the later implementation-dependent stages of system development, but it can also be used at the requirements specification level.

✓ *Assumption/Guarantee Style (A/G Style, Assumption/Committment Style)* – a component is specified in terms of an assumption and a guarantee, what means whenever input from the environment behaves in accordance with the assumption, the specified component is required to fulfill the guarantee.

✓ *Graphical Style* – based on tables and diagrams that are more readable and which can be schematically translated into equational or relational style (or into specifications expressed in predicate logic).[2]

It is also possible to use in FOCUS a combination of specification styles – one part of specification is expressed in one style and another part in another style. For example, we can specify a system in A/G style, where the assumption part is specified in the equational or in the relational style and the guarantee part is specified in the graphical style (see the FlexRay case study in Section 4.2).

1.3.4. Elementary and Composite Specifications

In FOCUS we can have both *elementary* and *composite* specifications. Elementary specifications (see Figure 1.1) are the atomic blocks for system representation, and in this case we distinguish among three frames, which correspond to the stream types in the specification: the *untimed*, the *timed* and the *time-synchronous frame* (see Section 1.3.1).

Any *elementary* FOCUS specification has the following syntax:

Name (Parameter_Declarations) ══════════════ Frame_Labels ══
in   *Input_Declarations*
out   *Output_Declarations*
Body

Name is the name of the specification; *Frame_Labels* lists a number of frame labels, e.g. *untimed*, *timed* or *time-synchronous*, that correspond to the stream

---

[2] There exists of course also a schematic translation in the other direction.

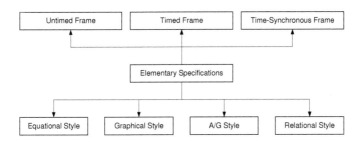

Figure 1.1.: FOCUS Specification Frames and Styles: Elementary Specifications

types in the specification (see Section 2.2); *Parameter_Declarations* lists a number of parameters (optional); *Input_Declarations* and *Output_Declarations* list the declarations of input and output channels respectively. *Body* characterizes the relation between the input and output streams, and can be a number of formulas, or a table, or diagram or a combination of them.

Having a timed specification we have always to deal with infinite timed streams by the definition of *semantics of timed frame*.

Definition: 1.1
The semantics of any elementary timed specification S (written $[\![S]\!]$) is defined in [BS01] to be the formula:

$$i_S \in I_S^{\frac{\infty}{S}} \ \wedge \ o_S \in O_S^{\frac{\infty}{S}} \ \wedge \ B_S \tag{1.1}$$

where i_S and o_S denote (lists of) input and output channel identifiers, I_S and O_S denote their corresponding types, and B_S is a formula in predicate logic that describes the body of the specification S. denotes the corresponding . □

To denote that the (lists of) input and output channel identifiers, I and O, build the syntactic interface of the specification S the following notation is used:

$$S \in (I \triangleright O)$$

The sets (lists) i_S and o_S of input and output channel identifiers of the specification S must be disjoint (see also Section 3.2).

For the most general case of specification we also need to argue about its parameters. For these purposes we extend the definition from [BS01] of the semantics of an elementary timed specification (Definition 1.1) to one of an elementary timed *parameterized* specification:

Definition: 1.2

For any elementary timed parameterized specification S we define its semantics, written $[\![S]\!]$, to be the formula:

$$i_S \in I_S^\infty \ \wedge \ p_S \in P_S \ \wedge \ o_S \in O_S^\infty \ \wedge \ B_S \qquad (1.2)$$

where i_S and o_S denote lists of input and output channel identifiers, I_S and O_S denote their corresponding types, p_S denotes the list of parameters and P_S denotes their types, B_S is a formula in predicate logic that describes the body of the specification S. □

Composite specifications (see Figure 1.2) are built hierarchically from elementary ones using constructors for composition and network description and can be represented in the *graphical*, the *constraint* and *operator* style. Semantics of a composite FOCUS specification is defined in [BS01] as given below.

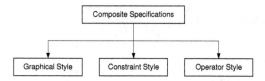

Figure 1.2.: FOCUS Specification Styles: Composite Specifications

Definition: 1.3

For any composite specification S consisting of n subspecifications $S_1, \ldots, S_n$, we define its semantics, written $[\![S]\!]$, to be the formula:

$$[\![S]\!] \stackrel{def}{=} \exists\, l_S \in L_S^\infty : \bigwedge_{j=1}^{n} [\![S_j]\!] \qquad (1.3)$$

where l_S denotes a list of local channel identifiers and L_S denotes their corresponding types. □

⚠ **Remark:** A component S with input channels $i_1, \ldots, i_n$, output channels $o_1, \ldots, o_m$ and parameters $p_1, \ldots, p_k$ (e.g. in some composite specification) can be referred in constraint style by

$$(o_1, \ldots, o_m) := S(p_1, \ldots, p_k)(i_1, \ldots, i_n)$$

⚠ **Remark:** One FOCUS stream can be an input of many components without using an extra component to split this stream, but to merge a number of streams into one stream some extra component is needed.

1.3.5. Sheaves and Replications

A specified system can contain a number of copies of channel of the same type or several instances of the same component. If this number of copies is finite, fixed and small enough to have a readable specification we can use the simple composition kinds. In the cases when

(1) the number of copies must be specified as some variable of type $\mathbb{N}$ or

(2) the number of copies is finite and fixed, but too large to have a readable system specification

the additional notions of *sheaf of channels* and *replication of specifications* are needed.

If a number of channels (streams) of the same type need to be represented in a specification, the concept of *sheaf of channels* [BS01] can be used. A sheaf of channels in FOCUS can be understood as an indexed set of channels. A system with the input sheaf of channels $x_1, \ldots, x_k$ (each of the channels has the type M_1), and the output sheaf of channels $y_1, \ldots, y_r$ (each of the channels has the type M_2) will be represented in FOCUS as the following specification S:

$$
\begin{array}{l}
\hline
\text{S} \hspace{6em} \text{Frame\_Labels} \\
\hline
\text{in} \quad x[\{1, \ldots, k\}] : M_1 \\
\text{out} \quad y[\{1, \ldots, r\}] : M_2 \\
\hline
Body \\
\hline
\end{array}
$$

The FOCUS technique for *replication of specifications* allows us to specify in simple and readable way a system that uses a number of several instances of the same component. Let discuss the FOCUS representation of the specification replication [BS01].

$$
\begin{array}{l}
\hline
\text{C(constant } p \in P) \hspace{4em} \text{Frame\_Labels} \\
\hline
\text{in} \quad i_1 : I_1; \; \ldots; \; i_n : I_n \\
\text{out} \quad o_1 : O_1; \; \ldots; \; o_m : O_m \\
\hline
Body \\
\hline
\end{array}
$$

Let Cid be a set of component identifiers and $g \in Cid \rightarrow P$. A network $RepC$ consisting of exactly one instance of the specification C for each identifier $j \in Cid$ is described in FOCUS by the expression

$$\otimes_{j \in Cid} C(g(j))(i_1[j], \ldots, i_n[j], o_1[j], \ldots, o_m[j]) \tag{1.4}$$

where $\otimes$ denotes in FOCUS the mutual feedback operator, which is defined as follows:

$$[\![S_1 \otimes S_2]\!] \stackrel{\text{def}}{=} \exists l \in L^\infty : [\![S_1]\!] \wedge [\![S_2]\!]$$

The syntactic interface of *RepC* is equal to the union of the external interfaces of its component specifications:

$$I_{RepC} = i_1[Cid]; \ldots; i_n[Cid]$$
$$O_{RepC} = o_1[Cid]; \ldots; o_m[Cid]$$

The graphical representation of the specification *RepC* is given below.

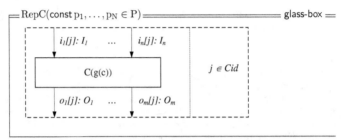

Because all inputs and outputs of *RepC* are sheaves of channels – n input and m output sheaves, and moreover, the parameters of the specification *RepC* also build a sheaf.

1.3.6. Refinement

System refinement provides a natural mechanism for structuring complex systems for increased readability. It is now widely recognized that it is not advisable and in most cases even impossible to make a concrete implementation of a large system from its abstract requirements specification in a single step. In practice a stepwise development is used – the requirements specification is refined into a concrete implementation stepwise, via a number of intermediate specifications (see [Bro97]).

Thus, a system development process has several levels of abstraction: from a requirement specification to a concrete implementation description. All these levels can be split into the following phases:

1. **Requirement phase**: Developing abstract formalization of an informal description. It will be used as the basis for the next phases.

2. **Design phase**: Developing the architecture of the system and refining it up to required level of granularity.

3. **Implementation phase**: Transformation of the specification, developed in the design phase, into one of the supplied implementation languages.

The presented here approach concentrates on the first two phases.

The FOCUS specification framework uses three basic refinement relations:

Behavioral Refinement:
> The related specifications S_1 and S_2 must have the same syntactic interface. The refined specification S_2 may meet further requirements in addition to the requirements on the more abstract specification S_1. At the same time, the more concrete specification S_2 must meet all the requirements on the specification S_1. This kind of refinement is used to reduce the number of possible output histories for a given input history. The formal definition of behavioral refinement [BS01] (also called *property refinement*) is presented in Section 3.1.2.

Interface Refinement:
> This kind of refinement is a generalization of the behavioral refinement – it allows to work on the different levels of interface abstraction: the related specifications S_1 and S_2 may have different syntactic interface. The interface refinement is used, e.g. when

> ✓ the datatypes of messages and channels is changed,

> ✓ one channel is represented by several channels or vice versa,

> ✓ one step of interaction is replaced by several steps of interaction or vice versa,

> ✓ additional error handling is included into a system specification, etc.

> The formal definition of interface refinement [BS01] is presented in Section 3.1.3.

Conditional Refinement:
> The conditional refinement is a generalization of the interface refinement – it allows the introduction of additional input assumptions. The formal definition of conditional refinement [BS01] is presented in Section 3.1.4.

We are using the definitions of refinement from [BS01].

Definition: 1.4
A specification S_2 is called a *behavioral refinement* $(S_1 \rightsquigarrow S_2)$ of a specification S_1 if

✓ they have the same syntactic interface and

✓ any I/O history of S_2 is also an I/O history of S_1.

The relation $\rightsquigarrow$ of behavioral refinement is defined (see also [BS01]) by equivalence

$$(S_1 \rightsquigarrow S_2) \iff (\llbracket S_2 \rrbracket \Rightarrow \llbracket S_1 \rrbracket) \tag{1.5}$$

□

Formally it means, that any I/O history of S_2 is an I/O history of S_1, but S_1 may have additional I/O histories.

Definition: 1.5

Let S_1, S_2, D, and U be specifications s.t.

$$S_1 \in (I_1 \triangleright O_1), \quad S_2 \in (I_2 \triangleright O_2), \quad D \in (I_1 \triangleright I_2), \quad U \in (O_2 \triangleright O_1)$$

The relation $\overset{(D,U)}{\rightsquigarrow}$ of interface refinement is defined as follows.

$$\begin{aligned}
& S_1 \overset{(D,U)}{\rightsquigarrow} S_2 \\
\Leftrightarrow\ & S_1 \rightsquigarrow (D \succ S_2 \succ U) \\
\Leftrightarrow\ & [\![\, D \succ S_2 \succ U \,]\!] \Rightarrow [\![\, S_1 \,]\!]
\end{aligned} \qquad (1.6)$$

where $\succ$ denotes in FOCUS piped composition of specifications. □

U and D are called *representation specifications*, and S_1 and S_2 the *abstract* and *concrete specification*, respectively (see Figure 1.3).

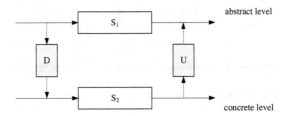

abstract level

concrete level

Figure 1.3.: Interface Refinement

Interface refinement is a generalization of behavioral refinement, and conditional refinement can be seen as generalization of interface refinement.

Definition: 1.6

Let S_1, S_2, D, U, and C be specifications s.t.

$$S_1 \in (I_1 \triangleright O_1), \quad S_2 \in (I_2 \triangleright O_2)$$
$$D \in (I_1 \triangleright I_2), \quad U \in (O_2 \triangleright O_1), \quad C \in (O_1 \triangleright I_1),$$

The relation $\overset{(D,U)}{\rightsquigarrow}_C$ of conditional interface refinement is defined as follows.

$$\begin{aligned}
& S_1 \overset{(D,U)}{\rightsquigarrow}_C S_2 \\
\Leftrightarrow\ & S_1 \rightsquigarrow_C (D \succ S_2 \succ U) \\
\Leftrightarrow\ & [\![\, C \,]\!] \wedge [\![\, D \succ S_2 \succ U \,]\!] \Rightarrow [\![\, S_1 \,]\!]
\end{aligned} \qquad (1.7)$$

□

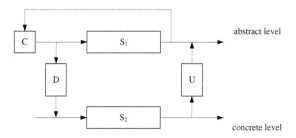

Figure 1.4.: Conditional Interface Refinement

Figure 1.4 illustrates the definition above. The specification C is called here *condition* and represents the additional input assumptions.

If D and U are identities, we speak about conditional *behavioral* refinement and write

$$S_1 \leadsto_C S_2$$

1.3.7. Causality

For a *timed* FOCUS specification S the relation $\mathcal{R}_S$ is called the I/O behavior of S:

$$\mathcal{R}_S \subseteq I_S{}^\infty \times O_S{}^\infty$$

We distinguish two kinds of causality for I/O behaviors:

✓ Weak causality

$$\forall\, x, y \in I^\infty;\ t \in \mathbb{N} : x{\downarrow}_t = y{\downarrow}_t \Rightarrow (\mathcal{R}.x){\downarrow}_t = (\mathcal{R}.y){\downarrow}_t \qquad (1.8)$$

✓ Strong causality

$$\forall\, x, y \in I^\infty;\ t \in \mathbb{N} : x{\downarrow}_t = y{\downarrow}_t \Rightarrow (\mathcal{R}.x){\downarrow}_{t+1} = (\mathcal{R}.y){\downarrow}_{t+1} \qquad (1.9)$$

where $\mathcal{R}.x$ and $(\mathcal{R}.x){\downarrow}_t$ are defined for any $x \in I_S{}^\infty$, $t \in \mathbb{N}$ and $\mathcal{R} \in I_S{}^\infty \times O_S{}^\infty$ by

$$\mathcal{R}.x \overset{\text{def}}{=} \{y \in O^\infty \mid (x, y) \in \mathcal{R}\}$$
$$(\mathcal{R}.x){\downarrow}_t \overset{\text{def}}{=} \{y{\downarrow}_t \in O^\infty \mid y \in \mathcal{R}.x\}$$

Specifying embedded real-time systems we always need (at least) weak causality. The strong causality is required to prevent Zeno paradoxes, if we want to have feedback loops (see Section 2.12.4).

1.4. JANUS

In this approach we use a highly simplified JANUS [Bro05] representation of the service notion. JANUS is a formal model of services and layered architectures that is based on the FOCUS theory of distributed systems. A JANUS service, like a FOCUS component, has a syntactic interface, but, in comparison to a component, a service has a partial behavior – it is defined only for a subset of its input histories.

Given a set M, by M^* and M^∞ will be denoted the set of finite and and the set of infinite sequences of elements of M respectively – both in JANUS and in FOCUS. In JANUS another definition for the timed streams as in FOCUS is introduced. The set of non-timed (finite and infinite) streams in JANUS will be denoted also like in FOCUS by

$$M^\omega \overset{\mathrm{def}}{=} M^* \cup M^\infty$$

A timed finite and infinite JANUS streams $M^{\underline{*}}$ and $M^{\underline{\infty}}$ over the set of messages M is defined respectively by the following functions

$$M^{\underline{*}} \overset{\mathrm{def}}{=} \bigcup_{n \in \mathbb{N}} ([1..n] \to M^*)$$
$$M^{\underline{\infty}} \overset{\mathrm{def}}{=} \mathbb{N} \to M^*$$
$$M^{\underline{\omega}} = M^{\underline{*}} \cup M^{\underline{\infty}}$$

Thus, in JANUS no $\sqrt{}$ are used and the time intervals are represented directly.[3] In general, there is nothing against usage the JANUS notations for the FOCUS specifications, because definitions have at the result very similar semantics. In our approach we define operators which cover the details of time interval representation. Thus, the results of "FOCUS on Isabelle" can be also extend to a complementary approach, "JANUS on Isabelle", that presents a coupling of a JANUS with Isabelle/HOL.

⚠ **Remark:** In contrast to FOCUS, the numeration of time intervals and of the elements in a sequence starts in JANUS (like in our Isabelle/HOL formalization of streams) from 0 and not from 1 (e.g. $\mathbb{N} \to M^*$ and not $\mathbb{N}_+ \to M^*$).

1.5. Outline

The thesis is organized as follows: Chapters 2 and 3 are the technical core of the thesis. Chapter 2 introduces a coupling and an implementation of the formal specification framework FOCUS in the generic theorem prover Isabelle/HOL – the translation of the elements of the FOCUS language – datatypes, streams (with comparison to other approaches in this area), operators, functions and predicates, different kinds of specification, and a number of FOCUS extras[4] –

---

[3] Our method to represent FOCUS (and JANUS) streams in Isabelle/HOL is equal to the JANUS method modulo Isabelle/HOL syntax.

[4] Sheaves of channels, specification replications, etc.

into Isabelle/HOL, i.e. deep embedding of FOCUS into Isabelle/HOL. In this chapter also are presented a number of syntax extensions for specification of timed systems. Chapter 3 presents the ideas of the specification and verification methodology, as well as the ideas of the so-called refinement-based verification. Chapter 4 shows feasibility of the approach on three case studies that cover different application areas and the different specification elements (ordered by their size):

✓ Steam Boiler System (process control),

✓ FlexRay communication protocol (data transmission),

✓ Automotive-Gateway System (memory and processing components, data transmission).

Chapter 5 summarizes the whole work.The Isabelle/HOL definitions and lemmas about the FOCUS operators are presented in Appendix A. Appendix B contains the Isabelle/HOL specifications and proofs for the case studies.

1. Introduction

2. FOCUS on Isabelle

This chapter introduces a coupling and an implementation of the formal specification framework FOCUS in the generic theorem prover Isabelle/HOL. We present here the translation of the FOCUS specifications of embedded real-time systems into Isabelle/HOL for proving properties of these systems.

The main questions discussed in this chapter are:

✓ How will the translation of FOCUS specifications and main concepts into Isabelle/HOL (deep embedding) be done?

✓ Is there anything special on the specification of *embedded real-time systems*?

✓ Which of the streams representation approaches is more appropriate for the case of FOCUS specifications of embedded real-time systems?

✓ Do we need some restrictions on specifications of such systems? Can we use this knowledge to simplify the specification in FOCUS and make it in methodological way?

✓ How will FOCUS datatypes be represented in Isabelle/HOL?

✓ How will the standard FOCUS operators be represented in Isabelle/HOL?

✓ Do we need some special operators or definitions for simplification of the FOCUS specification of the embedded real-time systems?

✓ Which kind of graphical specification techniques is especially appropriated in this case?

✓ How the semantics of a FOCUS specification be defined in Isabelle/HOL?

✓ Is there anything special on the specification with sheaves of channels or with replications?

✓ How the correctness of the relations between the sets of input, output and local channels can be proved? Can we define some proof schemata to make these proofs automatically?

✓ Are in FOCUS some kinds of specification constructions which are not very well situated to the translation to Isabelle/HOL? If this is the case, how we can reformulate them without changing their semantics?

2.1. Specification of Embedded Real-Time Systems

In this section we discuss the special features of the specification of embedded real-time systems: why the focus on the timed domain is such important and how the original FOCUS syntax can be extended to have more readable specifications of such kind of systems.

Specifying embedded real-time systems we always need to argue about time. The notion of time takes center stage for this kind of systems and abstracting from time we may loose very important properties, e.g. the causality property, that are not only very important for the system, but also help us to make proofs easier. Thus, the timed domain is the most important one for representation of distributed systems with real-time requirements. Therefore, the better way to represent a real-time system in FOCUS is to use timed specification, or, more precisely, specifications with timed or time-synchronous frame (see Section 1.3.4). All input, output and local streams in such specifications are timed or time-synchronous and, by the FOCUS definition of timed stream, infinite.

Specification of a real-time system in the untimed frame may be in some cases shorter or more elegant from mathematical point of view, but case studies have shown, that to understand such specifications and to argue about their properties is in many cases much more difficult in comparison to the corresponding specifications in the timed frame that use causality property explicitly. Moreover, abstraction from timing aspects can easily lead to specification mistakes because of difficulties of correct abstraction.

Hence, we can restrict the FOCUS specification domain for representation embedded real-time systems to only timed and time-synchronous systems. This simplifies the translation into Isabelle and also allows us to concentrate on the timing properties to have not only more clear and readable specifications, but also simpler proofs about them.

Considering causality (weak or strong) it is simpler and also more readable to argue not about single messages in a timed stream, but about a sequence of messages that are present in this stream at some time interval. This sequence can be in general empty, contain a single message or a number of messages. In the case of time-synchronous stream this sequence must always contain exactly one message.

For easier argumentation about the behavior of a component at some time interval we introduce in FOCUS a special kind of tables and a number of new operators (see Section 2.5).

The concrete meaning of a time interval is not defined in the FOCUS specification, but it must be specified additionally as a remark to the specification. Otherwise a number of questions can be obtained, e.g. the following ones: Do all time intervals have the same duration? If all time intervals have the same (constant) duration, how much seconds (nanoseconds, milliseconds, minutes, hours, etc.) does it take?

This interpretation flexibility allows to specify systems also for the case where the "time intervals" does not have the same (constant) duration and are understood as a formal technique for a causality representation.

2.2. Stream Representation

As mentioned in the previous section, the timed domain is the most important one for representation of distributed systems with real-time requirements. Therefore, we restrict our approach only to the timed specifications. With other words, to specify the behavior of a real-time system one simply needs to use *infinite timed streams* – to represent the input and the output streams.

The type of *finite timed streams* will be used only if some argumentation about a timed stream that was truncated at some point of time is needed.

The type of *finite untimed streams* will be used to argue about a sequence of messages that are transmitted during a time interval.

The type of *infinite untimed streams* will be used in the case of timed specifications only to represent local variables of FOCUS specifications (see also Section 2.7).

There are different ways of formalizing streams. They have different advantages and disadvantages. One of the ways to represent FOCUS streams is to use the coalgebraic approach (see [JR97]), but the representation of FOCUS streams in a coalgebraic domain (see [Spi03]) is more difficult to understand in practice as an inductive one in Isabelle/HOL, especially for the case of restriction the specification domain to only real-time systems.

The representation of FOCUS streams in Isabelle/HOLCF that was done by D. von Oheimb (see [vO05]) does not cover representation of FOCUS timed streams, which are the most important for the specification of embedded real-time systems. HOLCF (see [Reg94], [Reg95] and [MNvOS99]) is the definitional extension of Church's Higher-Order Logic with Scott's Logic for Computable Functions that has been implemented in Isabelle. HOLCF supports standard domain theory but also coinductive arguments about lazy datatypes. The main disadvantage of using HOLCF in practice is difficulty of logic understanding in comparison to HOL.

The further development of the FOCUS stream representation in Isabelle/HOLCF is presented by the approach of B. Gajanovic and B. Rumpe [GR06], that covers HOLCF specification of many important operators on streams, like concatenation, delete prefixes, take an element of the stream etc., as well as the properties of these operators. But this representation of the FOCUS streams in Isabelle/HOLCF covers only the general representation of streams, and abstracts from the representation of timing aspects as well as from the question how to deal with proofs for such translated specifications. Thus, the representation of the timing aspects can be done as an extension, but the resulting construction will be much more complicated than is needed for system specification in the embedded real-time domain.

To represent FOCUS streams in Isabelle/HOL we need to take into account both properties of FOCUS and Isabelle/HOL. In Isabelle/HOL we can represent streams in the two following ways. The first way is the representation of streams as

$$\alpha \ seq = \mathbb{N} \rightarrow \alpha \ option$$

where the datatype

$$\alpha \ option \equiv None \mid Some \ \alpha$$

and *None* denotes a non-existing element (see [Pau94] and [NS95]). This approach is claimed to be inconvenient in practice to prove equalities of arbitrary functions (see [DGM97]), because for every operation over such a stream the notion of stream normal form must be used explicitly to avoid the case in which *None* appears within a sequence the specification, but it is not straightforward to construct the normal form.

The second way, the representation of streams as the disjoint sum of finite stream (lists) and infinite streams (functions), has been chosen by C.-T. Chou and D. Peled (see [CP96]) and by S. Agerholm (see [Age94]). The main difficulties (see [DGM97]) in using this approach arise from type comparison in the definitions of stream processing functions – all inputs are finite, all inputs are infinite or some of them are finite and some infinite – several versions of function definitions are needed. But in the case we work only with timed frames we do not have this disadvantage, because we deal with timed streams that are always infinite[1]. Moreover, such a representation in this case leads to more clear specification structure. This representation is the most natural one to FOCUS and will be used in this approach for the our representation of FOCUS in Isabelle/HOL.

To represent a discrete notion of time we will use the JANUS representation of infinite timed streams: an infinite timed stream of some type M is represented as a function from natural numbers to the finite sequences (lists) of the type M that, where each finite sequence corresponds to the sequence of messages transmitted in the corresponding time interval. A finite timed stream of some type M is represented as a finite sequences of finite sequences (lists) of the type M, where again each finite sequence of the type M corresponds to the sequence of messages transmitted in the corresponding time interval.

The definition in Isabelle/HOL of corresponding types is given below:

✓ Finite timed streams of type *'a* are represented by the type *'a fstream*, which is an abbreviation for the type *'a list list*.

✓ Finite untimed streams of type *'a* are represented by the list type: *'a list*.

✓ Infinite timed streams of type *'a* are represented by the type *'a istream*, which represents the functional type *nat* $\Rightarrow$ *'a list*.

✓ Infinite untimed streams of type *'a* are represented by the functional type *nat* $\Rightarrow$ *'a*.

datatype *'a stream* = *FinT 'a fstream*
      | *FinU 'a list*
      | *InfT 'a istream*
      | *InfU nat* $\Rightarrow$ *'a*

---

[1] The timed streams must be infinite because time never halts.

⚠ **Remark:** Translating definitions about finite untimed streams (lists) we will use predefined Isabelle/HOL theory List.thy. This theory is easy to use and it has a number of proved properties about lists that can be useful for making proofs. We also define a number of additional lemmas about lists that were useful in the case studies (see Chapter 4) as the theory ListExtras.thy (see Appendix A.5).

A special kind of streams are *time-synchronous streams*, where each message represents one time interval (such kind of stream does not contain $\sqrt{}$). In FOCUS these streams are represented as untimed streams given a particular timed interpretation, i.e. their semantics is timed, but the syntax is equal to the syntax of untimed streams. Strictly speaking, the FOCUS time-synchronous streams cannot be combined in a composite specification with any other kind of streams, neither with untimed streams, because of different semantics, nor with timed streams, because of different syntax.

For this reasons, such a stream is modeled in this approach as a timed stream for which a predicate ts [2] holds (for more detail see Section 2.10). This representation allows to combine timed and time-synchronous specification.

⚠ **Remark:** By the FOCUS definition both timed and time-synchronous streams are infinite. It is important to note, that if we make a time-synchronous stream untimed we get always an infinite stream, but making a timed stream untimed we can get in some cases a finite untimed stream as result.

⚠ **Remark:** The numeration of stream elements starts in our Isabelle/HOL representation from 0 because of the types definition in Isabelle/HOL, where in FOCUS it starts from 1. This is taken into account in the representation of the FOCUS operators in Isabelle/HOL. However, arguing about timed streams via time intervals, e.g. using the operator ti (see Section 2.5.3) or a tiTable (see Section 2.6), we avoid this problem automatically.

In [BS01] a large set of the operators on streams is defined. We have represented the main part of them, which is needed to argue about real-time systems, in Isabelle/HOL. We discuss here only the operators that are presented in Section 1.3.2. We present only the signatures of another operators – the whole representation is give in Appendix A as an Isabelle/HOL theory.

⚠ **Remark:** $[\![\, E \,]\!]_{Isab}$ denoted here the Isabelle/HOL representation of the FOCUS expression E.

---

[2]The predicate $ts(s)$ is true, if every list of infinite stream s contains exactly one element (see Section 1.3.2).

2.3. Representation of Datatypes

Before we start to discuss the representation of the Focus operators in Isabelle/HOL, we need to illuminate how the Focus datatypes will be represented. Table 2.1 summarises the translation of Focus datatypes into Isabelle/HOL. As mentioned in Section 1.2, the base types in Isabelle/HOL are *bool*, the type of truth values, and *nat*, the type of natural numbers. The base type constructors are *list*, the type of lists, and *set*, the type of sets. The polymorphic types can be defined using *type variables* denoted by $'a$, $'b$ etc.

An enumeration type can be represented in Focus in two ways that have the same semantics:

> **type** $T = e_1 \mid \cdots \mid e_n$ and
>
> **type** $T = \{e_1, \ldots, e_n\}$

We represent them in Isabelle/HOL by

> *datatype* $T = e_1 \mid \cdots \mid e_n$

The Focus *records type RV*

> **type** $RV = con_1(sel_1^1 \in T_1^1, \ldots, sel_{k_1}^1 \in T_{k_1}^1)$
>
> $\cdots$
>
> $\mid con_n(sel_1^n \in T_1^n, \ldots, sel_{k_n}^n \in T_{k_n}^n)$

will be translated into Isabelle/HOL as follows:

> *datatype*
> $RV = con_1 \ con\_type_1 \mid \ldots \mid con_n \ con\_type_n$

where con_i is a constructor name and the datatype con_type_i, $1 \leq i \leq n$, is defined as follows

> *record* $con\_type_i =$
> $sel_1^i :: [\![T_1^i]\!]_{Isab}$
> $\cdots$
> $sel_{k_i}^i :: [\![T_{k_i}^i]\!]_{Isab}$

If T_i is a composed type, then its Isabelle/HOL semantics must be written in " ". If some type consists only of one record, i.e. $n = 1$, we can represent them in Isabelle/HOL simply using only a record-definition. For example, the Isabelle/HOL semantics of the Focus type

> **type** $Config = conf \ (schedule : \mathbb{N}^*, cycleLength : \mathbb{N})$

is specified by

> *record* $Config =$
> $schedule :: \text{"nat list"}$
> $cycleLength :: nat;$

The Isabelle/HOL semantics of an undefined FOCUS type M will be some type variable ($'a$, $'b$ etc.). Thus, if a record type R is based on some undefined type T_i (is *polymorphic*) and $[\![T_i]\!]_{Isab} = {}'a$, then $'a$ will be added to the Isabelle/HOL definition of R between the keyword **record** and the name of type, R.

For example, let the datatype *Data* be undefined. Then the following FOCUS type is polymorphic

type $Message = msg$ ($message\_id : \mathbb{N}, ftcdata : Data$)

and must be represented in Isabelle/HOL by

record $'a$ Message =
   message_id :: nat
   ftcdata :: $'a$;

FOCUS type, T	Isabelle/HOL representation, $[\![T]\!]_{Isab}$
$\mathbb{N}$	*nat*
$\mathbb{B}ool$	*bool*
M^*	$[\![M]\!]_{Isab}$ *list*
$M^{\underline{*}}$	$[\![M]\!]_{Isab}$ *fstream*
M^∞	*nat* $\Rightarrow [\![M]\!]_{Isab}$
$M^{\underline{\infty}}$	$[\![M]\!]_{Isab}$ *istream*
M^ω	$[\![M]\!]_{Isab}$ *stream*
$x \in M$	$([\![x]\!]_{Isab} :: [\![M]\!]_{Isab})$
type $T = e_1 \mid \cdots \mid e_n$	***datatype** $T = e_1 \mid \cdots \mid e_n$*
type $T = \{e_1, \ldots, e_n\}$	***datatype** $T = e_1 \mid \cdots \mid e_n$*
	***record** $R =$*
	$sel_1 :: [\![T_1]\!]_{Isab}$
type $R =$	
$constr(sel_1 \in T_1, \ldots, sel_k \in T_k)$	$\cdots$
	$sel_k :: [\![T_k]\!]_{Isab}$

Table 2.1.: Isabelle/HOL representation of the FOCUS types (M is here some datatype)

2.4. FOCUS Operators Representation

In this section we discuss the representation of FOCUS operators (presented in Section 1.3.2) in Isabelle/HOL. The new defined FOCUS operators as well as their representation in Isabelle/HOL will be given in Section 2.5. Isabelle/HOL specifications of all these operators are presented in Appendices A.1, A.2 and A.3. From here, we add the prefix *fin_* to all Isabelle/HOL operators defined on finite timed streams, and the prefix *inf_* to all Isabelle/HOL operators defined on infinite timed streams.

2.4.1. nth Message of a Stream

This operator is more appropriate for untimed FOCUS streams. To use this operator for timed streams is unusual, but we introduce it for all kind of streams. For a finite untimed stream s we can use the Isabelle/HOL operator *nth*, predefined in the theory *List.thy*[3]:

$$[\![s.(n+1)]\!]_{Isab} \equiv nth \; s \; n$$

We can take the *nth* message of an infinite untimed stream s "directly":

$$[\![s.(n+1)]\!]_{Isab} \equiv s \; n$$

For finite and infinite timed streams we define operators *fin_nth*, $[\![s.n]\!]_{Isab} \equiv$ *(fin_nth s n)*, and *inf_nth*, $[\![s.n]\!]_{Isab} \equiv$ *(inf_nth s n)*, respectively.

consts
    fin_nth :: $'a \; fstream \Rightarrow nat \Rightarrow 'a$
    inf_nth :: $'a \; istream \Rightarrow nat \Rightarrow 'a$

primrec
fin_nth_Cons:
  fin_nth $(hds \; \# \; tls) \; k =$
    $(\; if \; hds = []$
      $then \; fin\_nth \; tls \; k$
      $else \; (\; if \; (k < (length \; hds))$
          $then \; nth \; hds \; k$
          $else \; fin\_nth \; tls \; (k - length \; hds) \;))$
primrec
inf_nth $s \; 0 =$
    $hd \; (s \; (LEAST \; i.(s \; i) \neq []))$
inf_nth $s \; (Suc \; k) =$
    $(\; if \; ((Suc \; k) < (length \; (s \; 0)))$
      $then \; (nth \; (s \; 0) \; (Suc \; k))$
      $else \; (\; if \; (s \; 0) = []$
          $then \; (inf\_nth \; (inf\_tl \; (inf\_drop$
            $(LEAST \; i. \; (s \; i) \neq []) \; s)) \; k \;)$
          $else \; inf\_nth \; (inf\_tl \; s) \; k \;))$

---

[3]The *nth* operator is defined in Isabelle/HOL only for nonempty streams: the definition of this operator is given by recursion on the list, and the base case (the list is an empty one) of the recursion is omitted, see [Nip05].

2.4.2. Length of a Stream

The length of an infinite untimed stream is by definition equal to ∞. It is insufficient to use the length operator for infinite timed streams, because this operator counts a number of messages in a stream, thus, it argues over single messages of a stream and not over its time intervals. In general, the length of an infinite timed stream (i.e. a number of messages in this stream) can also be finite. The notion that a timed stream has infinitely many messages can be expressed by the property "the stream has infinitely many nonempty time intervals" (see Section 2.5.3).

For a finite untimed stream s we can use the Isabelle/HOL operator *length*, predefined in the theory *List.thy*:

$$[\![\#s]\!]_{Isab} = length \; [\![s]\!]_{Isab}$$

For a finite timed stream we define operator *fin_length* that counts the number of messages of all time intervals of the stream:

```
consts
  fin_length :: 'a fstream ⇒ nat
primrec
  fin_length [] = 0
  fin_length (x#xs) = (length x) + (fin_length xs)
```

According to this definition, we get for an finite timed stream s

$$[\![\#s]\!]_{Isab} = fin\_length \; [\![s]\!]_{Isab}$$

2.4.3. Concatenation Operator

We represent the concatenation operator on streams x and y, $x \frown y$, according to the types of x and y in Table 2.4.3. This operator is defined only if both streams are timed or both streams are untimed, the mix version makes no sense. If both streams are finite, we use the function @ that is predefined in the Isabelle/HOL theory *List.thy* as concatenation function for two lists. If the first stream is finite, and the second stream is infinite, we have as result an infinite stream generated by a function *fin_inf_append*. If the first stream is infinite, we have as result this stream, otherwise we append the second stream to the first one.

We define the function *fin_inf_append* as follows:

```
constdefs
  fin_inf_append :: 'a list ⇒ (nat ⇒ 'a) ⇒ (nat ⇒ 'a)
  fin_inf_append us s ≡
    (λ i. ( if (i < (length us))
            then (nth us i)
            else s (i − (length us)) ))
```

type of x	type of y	$[\![x \frown y]\!]_{Isab}$
finite timed	finite timed	$[\![x]\!]_{Isab}$ @ $[\![y]\!]_{Isab}$
finite timed	finite untimed	–
finite timed	infinite timed	$fin\_inf\_append\ [\![x]\!]_{Isab}\ [\![y]\!]_{Isab}$
finite timed	infinite untimed	–
finite untimed	finite timed	–
finite untimed	finite untimed	$[\![x]\!]_{Isab}$ @ $[\![y]\!]_{Isab}$
finite untimed	infinite timed	–
finite untimed	infinite untimed	$fin\_inf\_append\ [\![x]\!]_{Isab}\ [\![y]\!]_{Isab}$
infinite timed	finite timed	$[\![x]\!]_{Isab}$
infinite timed	finite untimed	–
infinite timed	infinite timed	$[\![x]\!]_{Isab}$
infinite timed	infinite untimed	–
infinite untimed	finite timed	–
infinite untimed	finite untimed	$[\![x]\!]_{Isab}$
infinite untimed	infinite timed	–
infinite untimed	infinite untimed	$[\![x]\!]_{Isab}$

Table 2.2.: Isabelle/HOL representation of the prefix ordering on streams

For the concatenation operator we have proved a number of lemmas (see the Isabelle/HOL theory *stream.thy* in Appendix A.1), e.g.:

$$\#x, \#y \in \mathbb{N}: \ \#x \frown y = \#x + \#y$$

$$\langle\rangle \frown z = z$$

$$s_1 = \langle x \rangle \frown s \Rightarrow \mathsf{ti}(s_1, i + 1) = \mathsf{ti}(s, i)$$

$$\#x, \#y \in \mathbb{N}, \ \#z = \infty: \ x \frown (y \frown z) = (x \frown y) \frown z$$

⚠ **Remark:** The following equality holds: $m \,\&\, s \equiv \langle m \rangle \frown s$

2.4.4. Prefix of a Stream

If the stream x is infinite, the predicate $x \sqsubseteq y$ will be *True* only if the stream y is infinite and equal to the first one. If both streams are finite, we can use the predicate $\leq$ predefined in the Isabelle theory *List.thy* as the prefix predicate for lists. For the case if the first stream is finite and the second one is infinite we define an Isabelle/HOL function *inf_prefix*. The prefix ordering on streams x and y, $x \sqsubseteq y$, is represented according to the types of the streams in Table 2.4.4.

```
consts
   inf_prefix :: 'a list ⇒ (nat ⇒ 'a) ⇒ nat ⇒ bool
primrec
   inf_prefix [] s k = True
   inf_prefix (x#xs) s k =
      ( (x = (s k)) ∧ (inf_prefix xs s (Suc k)) )
```

type of x	type of y	$[\![x \sqsubseteq y]\!]_{Isab}$
finite timed	finite timed	$[\![x]\!]_{Isab} \leq [\![y]\!]_{Isab}$
finite timed	finite untimed	*False*
finite timed	infinite timed	*inf_prefix* $[\![x]\!]_{Isab}$ $[\![y]\!]_{Isab}$ 0
finite timed	infinite untimed	*False*
finite untimed	finite timed	*False*
finite untimed	finite untimed	$[\![x]\!]_{Isab} \leq [\![y]\!]_{Isab}$
finite untimed	infinite timed	*False*
finite untimed	infinite untimed	*inf_prefix* $[\![x]\!]_{Isab}$ $[\![y]\!]_{Isab}$ 0
infinite timed	finite timed	*False*
infinite timed	finite untimed	*False*
infinite timed	infinite timed	$(\forall\ i.\ [\![x]\!]_{Isab}\ i = [\![y]\!]_{Isab}\ i)$
infinite timed	infinite untimed	*False*
infinite untimed	finite timed	*False*
infinite untimed	finite untimed	*False*
infinite untimed	infinite timed	*False*
infinite untimed	infinite untimed	$(\forall\ i.\ [\![x]\!]_{Isab}\ i = [\![y]\!]_{Isab}\ i)$

Table 2.3.: Isabelle/HOL representation of the prefix ordering on streams

2.4.5. Truncate a Stream

The operator $s \downarrow_t$ is defined in FOCUS to truncate a timed infinite stream s at a certain point t in time. We define the corresponding Isabelle/HOL operator first only for $t \in \mathbb{N}$, i.e. for $t \neq \infty$:

consts
  inf_truncate :: $(nat \Rightarrow {}'a) \Rightarrow nat \Rightarrow {}'a\ list$
primrec
  inf_truncate s 0 $=$ [*s 0*]
  inf_truncate s (Suc k) $=$ (*inf_truncate s k*) @ [*s (Suc k)*]

For the case t can be also equal to ∞, we define another Isabelle/HOL operator, *inf_truncate_plus* (the Isabelle/HOL type *inat* denotes the type of natural numbers extended by a special value ∞ indicating infinity).[4] If the stream is of type M, the result will be of type M^{ω}.

constdefs
  inf_truncate_plus :: $'a\ istream \Rightarrow inat \Rightarrow {}'a\ stream$
  infT_truncate_plus s n
  $\equiv$
  case n of $(Fin\ i) \Rightarrow FinT$ (*inf_truncate s i*)
      $|\ \infty\ \ \Rightarrow InfT\ s$

2.4.6. Domain and Range of a Stream

It is insufficient to use the domain operator (see Section 1.3.2) for timed streams, because the argumentation will be not over time interval of a stream, but over single messages. Thus, we present here the Focus domain operator in Isabelle/HOL only for untimed, finite and infinite, streams.

(1) Domain of a finite untimed stream will be defined in two ways: the function *finU_dom s* returns the domain of the stream s as subset of *nat*, and the function *finU_dom_inat s* returns the domain of the stream s as subset of *inat*.

consts
  finU_dom :: $'a\ list \Rightarrow nat\ set$
primrec
  finU_dom [] $= \{\}$
  finU_dom $(x\#xs) = \{length\ xs\} \cup (finU\_dom\ xs)$

constdefs
  finU_dom_inat :: $'a\ list \Rightarrow inat\ set$
  finU_dom_inat s $\equiv \{x.\ \exists\ i.\ x = (Fin\ i) \land i < (length\ s)\}$

We define here the domain operator for finite untimed streams (Isabelle/HOL lists) recursively, because the main part of the Isabelle/HOL function over lists is defined in this way. However, for the *funU_dom* operator the following property holds:

  funU_dom xs $= \{i.\ i < length\ xs\}$

---

[4]This type is defined in the Isabelle/HOL theory *Nat_Infinity.thy* (see [vO]).

(2) Domain of an infinite untimed stream (the whole set of natural numbers extended by a special value indicating infinity):

constdefs
  infU_dom :: inat set
  infU_dom ≡ {x. ∃ i. x = (Fin i)} ∪ {∞}

The FOCUS range operator (see Section 1.3.2) will be represented in Isabelle/HOL for all kinds of streams.

(1) Range of a finite timed stream:

consts
  finT_range :: 'a fstream ⇒ 'a set
primrec
  finT_range [] = {}
  finT_range (x#xs) = (set x) ∪ finT_range xs

(2) Range of a finite untimed stream:

constdefs
  finU_range :: 'a list ⇒ 'a set
  finU_range x ≡ set x

(3) Range of an infinite timed stream:

constdefs
  infT_range :: 'a istream ⇒ 'a set
  infT_range s ≡ {y. ∃ i::nat. y mem (s i)}

(4) Range of an infinite untimed stream:

constdefs
  infU_range :: (nat ⇒ 'a) ⇒ 'a set
  infU_range s ≡ { y. ∃ i::nat. y = (s i) }

2.4.7. Time-synchronous Stream

We represent the FOCUS predicate **ts** $\in M^{\infty} \rightarrow \mathbb{B}$ool in Isabelle/HOL as follows:

constdefs
  ts :: 'a istream ⇒ bool
  ts s ≡ ∀ i. (length (s i) = 1)

We have proved a number of lemmas about properties of the predicate *ts* and its correlation with other operators (for the proofs see the Isabelle/HOL theory *stream.thy* in Appendix A.1), e.g.:

$$\forall\, x:\ \mathsf{ts}(x) \rightarrow (\forall\, t \in \mathbb{N}:\ \mathsf{ft.ti}(x,t) = \overline{x}.(t+1)) \tag{2.1}$$

2.4.8. Make Untimed

We define the operator $\overline{s}$ in Isabelle/HOL as follows. For the case of a finite (timed) stream s we use a function *fin_make_untimed*. For the case of an infinite (timed) stream s we have two subcases: if from some time interval all time intervals are empty, we truncate the stream on this time by the function *inf_truncate* (see Section 2.4.5), and use the function *fin_make_untimed*, otherwise we use the function *inf_make_untimed* (in this case the output stream still be an infinite one).

The function *fin_make_untimed* converts the finite timed stream to the corresponding finite untimed stream – the resulting stream will be represented by a single list of all messages. The function *inf_make_untimed* converts the infinite timed stream to infinite untimed one – it is assumed that the input stream has no infinite sequence of empty lists. The Isabelle/HOL definitions of the functions *fin_make_untimed* and *inf_make_untimed* are presented in the theory *stream.thy* (see Appendix A.1). The Isabelle/HOL function *make_untimed* gives an example how such kind of functions can be used, if we have no information about finiteness of a stream:

constdefs
```
    make_untimed :: 'a stream  ⇒ 'a stream
    make_untimed s ≡
        case s of (FinT x) ⇒ FinU (fin_make_untimed x)
              | (FinU x) ⇒ FinU x
              | (InfT x) ⇒
                  (if (∃ i.∀ j. i < j ⟶ (x j) = [])
                   then FinU (fin_make_untimed (inf_truncate x
                           (LEAST i.∀ j. i < j ⟶ (x j) = [])))
                   else InfU (inf_make_untimed x))
              | (InfU x) ⇒ InfU x
```

2.4.9. Time Stamp Operator

The Focus time stamp operator has the signature

$$\mathsf{tm} \in M^{\underline{\omega}} \times \mathbb{N}_+ \rightarrow \mathbb{N}$$

and yields for a timed stream s and a natural number k the index of time interval in which the kth message in the stream s is transmitted. This operator is defined in Focus only for such numbers k that $k \in \mathsf{dom}.\overline{s}$.

To represent this operator for infinite timed streams in Isabelle/HOL, we have defined the function *inf_tm* recursively on the number of messages using **recdef** kind of definition (the recursion goes here not uniformly – the primitive recursion does not work and an additional auxiliary lemma about termination is needed).

consts
  fin_tm :: *'a fstream* ⇒ *nat* ⇒ *nat*
primrec
  fin_tm [] *k* = *k*
  fin_tm (*x*#*xs*) *k* =
    (*if k* = *0*
    then 0
    else (*if* (*k* ≤ *length x*)
        then (*Suc 0*)
        else Suc(*fin_tm xs* (*k* − *length x*))))

consts
  inf_tm :: ('*a istream* × *nat*) ⇒ *nat*
recdef *inf_tm measure*(λ(*s*,*n*). *n*)
  inf_tm (*s*, *0*) = *0*

  inf_tm (*s*, *Suc i*) =
  (*if* (∀ *j*. *s j* = [])
    then 0
    else
      (*let*
        k = (*LEAST x WRT* (λ*n*. *n*). *s x* ≠ [])
      in
        (*if* (*Suc i*) ≤ (*length*(*s k*))
        then (*Suc k*)
        else (*let*
            i2 = (*Suc i*) − (*length* (*s k*));
            s2 = *inf_drop* (*Suc k*) *s*
          in
            inf_tm (*s2*, *i2*))
))
) (**hints** *intro*: *inf_tm_hint* [*rule_format*])

2.4.10. Filtering Operator

We define the filter function $M \circledS s$ in Isabelle/HOL as follows. If the stream s is a finite untimed one, we use the function *filter* predefined in the Isabelle/HOL theory *List.thy*:

$$[\![M \circledS s]\!]_{Isab} \equiv filter\ (\lambda\ y.y \in [\![M]\!]_{Isab})\ [\![s]\!]_{Isab}$$

If the stream s is an infinite untimed stream one, we use the function *filter_inf* defined in the Isabelle/HOL theory *Filter.thy* (see [Ber05]):

$$[\![M \circledS s]\!]_{Isab} \equiv filter\_inf\ (\lambda\ y.y \in [\![M]\!]_{Isab})\ [\![s]\!]_{Isab}$$

For the cases when s is a timed stream, finite or infinite, we define the corresponding functions *finT_filter* and *infT_filter*, which apply the *filter* function to each time interval. Thus, the filtering operator is defined for the timed streams in such way, that the time ticks $\sqrt{}$ are ignored.

constdefs
    finT_filter :: *'a set => 'a fstream => 'a fstream*
  finT_filter m s $\equiv$ *map* (λ *s. filter* (λ *y. y* $\in$ *m*) *s*) *s*

    infT_filter :: *'a set => 'a istream => 'a istream*
  infT_filter m s $\equiv$ (λi.(*filter* (λ *x. x* $\in$ *m*) (*s i*)))

The function *map* is predefined in the Isabelle theory *List.thy* – it applies a function that is the first argument of the *map* to the list that is its second argument.

2.4.11. Application Operator

The application operator map is defined in FOCUS only for untimed streams. For a finite stream s we represent this operator by the Isabelle/HOL function *map* predefined in the theory *List.thy*:

$$[\![\mathsf{map}(s,f)]\!]_{Isab} = map \ [\![f]\!]_{Isab} \ [\![s]\!]_{Isab}$$

For the case of an infinite stream s we use the predefined composition function $\circ$:

$$[\![\mathsf{map}(s,f)]\!]_{Isab} = [\![f]\!]_{Isab} \ \circ \ [\![s]\!]_{Isab}$$

⚠ **Remark:** If the function f is defined over M^* (i.e. over lists of messages) we can also use the operator $\mathsf{map}(s,f)$ for a *timed* stream s.

2.4.12. Stuttering Removal Operator

The stuttering removal operator $\propto$ is defined in FOCUS only for untimed streams. For the case of finite untimed streams we will use an Isabelle/HOL function *remdups*, which removes all duplications from the list (see the Isabelle/HOL theory *List.thy*):

$$[\![\propto .s]\!]_{Isab} = remdups \ [\![s]\!]_{Isab}$$

Having this restriction we do not use the infinite untimed streams in FOCUS specifications directly, but use the corresponding construction in Isabelle/HOL to represent local and state variables (see Appendix 2.7). Thus, the situation where we can use the stuttering removal operator $\propto$ to an infinite untimed stream is quite uncommon, but possible. For the case of infinite untimed streams we define the function *inf_remdups*:

$$[\![\propto .s]\!]_{Isab} = inf\_remdups \ [\![s]\!]_{Isab}$$

The definition of the function *inf_remdups* is kind of complicated (see Section A.1): the situation

$$\exists i: \ \forall j > i: \ s.i = s.j$$

must be taking into account. Thus, the stream $\propto .s$ can be also a finite one.

2.5. FOCUS Operators: Extensions

In this section we introduce a number of new FOCUS operators for the argumentation over time intervals as well as their representation in Isabelle/HOL.

2.5.1. Number of Time Intervals in a Finite Stream

We define the operator

$$\mathsf{Nti} \in M^{\underline{*}} \to \mathbb{N}$$

to denote the number of time intervals in a finite timed stream as follows:

$$\mathsf{Nti}(s) \stackrel{def}{=} \begin{cases} \#(\{\sqrt{}\} \, \textcircled{s} \, s) & \text{if } s.\#s = \sqrt{} \\ \#(\{\sqrt{}\} \, \textcircled{s} \, s) + 1 & \text{otherwise} \end{cases}$$

For example,

$$\mathsf{Nti}(\langle a_1, a_2, \sqrt{}, a_3, \sqrt{} \rangle) = 2$$
$$\mathsf{Nti}(\langle a_1, a_2, \sqrt{}, a_3, \sqrt{}, a_4 \rangle) = 3$$
$$\mathsf{Nti}(\langle a_1, a_2, \sqrt{}, a_3, \sqrt{}, a_4, \sqrt{} \rangle) = 3$$
$$\mathsf{Nti}(\langle a_1 \rangle) = 1$$
$$\mathsf{Nti}(\langle \rangle) = 0$$

The Isabelle/HOL definition of this operator is very simple, we just need to count the number of lists representing time intervals of the stream:

$$[\![\mathsf{Nti}(s)]\!]_{Isab} \equiv length \ [\![s]\!]_{Isab}$$

Using our Isabelle/HOL representation of timed stream, we do not need to distinguish between the cases "is $\sqrt{}$ the last message of the stream" or not, because in this representation, e.g., the following holds

$$[\![\langle a_1, a_2, \sqrt{}, a_3, \sqrt{}, a_4, \sqrt{} \rangle]\!]_{Isab}$$
$$\equiv \ [\![\langle a_1, a_2, \sqrt{}, a_3, \sqrt{}, a_4 \rangle]\!]_{Isab}$$
$$\equiv \ [\ [\ [\![a_1]\!]_{Isab}, [\![a_2]\!]_{Isab} \ ], \ [\ [\![a_3]\!]_{Isab} \ ], \ [\ [\![a_4]\!]_{Isab} \ ] \ ]$$

2.5.2. Timed Truncation Operator

The original FOCUS operator $s \downarrow_t$ is defined in [BS01] only for a timed infinite stream. We extend this operator also to finite timed streams:

$$s\downarrow_n \stackrel{def}{=} \begin{cases} s & \text{if } n = \infty \\ \langle \rangle & \text{if } n = 0 \\ r & \text{otherwise, where} \\ & \quad \text{if } \#s = \infty \vee (\#s \neq \infty \wedge (n < \mathsf{Nti}(s) \vee n = \mathsf{Nti}(s) \wedge s.\#s = \sqrt{})) \\ & \quad \text{then } r \sqsubseteq s \wedge \#\{\sqrt{}\} \textcircled{s} r = n \wedge r.\#r = \sqrt{} \\ & \quad \text{else } r = s \\ & \quad \text{fi} \end{cases}$$

Now, the last message in the truncated stream $s{\downarrow}_n$ is not a tick $\sqrt{}$ not only if s is empty, but also if $\mathsf{Nti}(s) < n$, and also if $\mathsf{Nti}(s) = s$ and the last entity of s is not a $\sqrt{}$.

We define the corresponding Isabelle/HOL operator first only for $t \in \mathbb{N}$, i.e. $t \neq \infty$:

```
consts
    fin_truncate :: 'a list ⇒ nat ⇒ 'a list
primrec
  fin_truncate [] n = []
  fin_truncate (x#xs) i =
     (case i of 0 ⇒ []
        | (Suc n) ⇒ x # (fin_truncate xs n))
```

For the case t can be also equal to ∞, we define another Isabelle/HOL operator, *fin_truncate_plus*:

```
constdefs
    fin_truncate_plus :: 'a list ⇒ inat ⇒ 'a list
  fin_truncate_plus s n
     ≡
   case n of (Fin i) ⇒ fin_truncate s i
        | ∞      ⇒ s
```

2.5.3. Time Interval

The operator $\mathsf{ti}(s,n)$ denotes the sequence of messages that are present on the channel s at the time interval between nth and $(n + 1)$th ticks:

$$\mathsf{ti} \in M^{\underline{\omega}} \times \mathbb{N} \to M^{*}$$

We define this operator in the Focus syntax as follows:

$$\mathsf{ti}(s, n) \stackrel{\mathrm{def}}{=} r, \ \ where \ \ s{\downarrow}_{n+1} = s{\downarrow}_n {}^\frown r {}^\frown \langle\sqrt{}\rangle \tag{2.2}$$

For example, let

$$s = \langle m_1, \sqrt{}, m_2, m_3, \sqrt{}, m_4, m_5, m_6, \sqrt{}, m_7, \sqrt{} \ldots \rangle,$$

then

$$\mathsf{ti}(s, 0) = \langle m_1 \rangle$$
$$\mathsf{ti}(s, 1) = \langle m_2, m_3 \rangle$$
$$\mathsf{ti}(s, 2) = \langle m_4, m_5, m_6 \rangle$$

$\ldots$

⚠ **Remark:** The FOCUS operator $\mathsf{ti}(s,n)$ corresponds to the JANUS "." operator: we can define the operator $\mathsf{ti}(s,n)$ in the JANUS syntax by

$$\mathsf{ti}(s,n) \stackrel{\text{def}}{=} s.(n+1)$$

In Isabelle/HOL we represent the operator $\mathsf{ti}(s,n)$ as follows:

✓ for the case of a finite stream s, we define operator ti as extended version of Isabelle/HOL operator nth for lists, $[\![\mathsf{ti}(s,t)]\!]_{Isab} \equiv ti\ [\![s]\!]_{Isab}\ [\![t]\!]_{Isab}$:

```
constdefs
  ti :: 'a fstream ⇒ nat ⇒ 'a list
ti s i ≡
  (if s = []
  then []
  else (nth s i))
```

✓ if s is an infinite stream, which is specified in Isabelle/HOL as mapping from natural numbers to lists, we can simply use expression $s\ n$: $[\![\mathsf{ti}(s,t)]\!]_{Isab} \equiv [\![s]\!]_{Isab}\ [\![t]\!]_{Isab}$.

2.5.4. Drop a Stream

The operator $s \uparrow_n$ denotes in JANUS the timed stream s without first n time intervals:

$$\uparrow\ \in M^{\underline{\omega}} \to \mathbb{N} \to M^{\underline{\omega}}$$

We define this operator for timed streams in FOCUS as follows:

$$\forall\, t \in \mathbb{N}:\ \mathsf{ti}(s \uparrow_k, t) = \mathsf{ti}(s, t+k) \tag{2.3}$$

We can also extend this operator in FOCUS to the operation with untimed streams:

$$\begin{aligned}
\langle\rangle \uparrow_k &= \langle\rangle \\
(\langle x\rangle \frown s) \uparrow_k &= \text{if } k = 0 \text{ then } \langle x\rangle \frown s \text{ else } s \uparrow_{k-1} \text{ fi}
\end{aligned} \tag{2.4}$$

To represent the drop operator for finite streams in Isabelle/HOL we can use the Isabelle/HOL operator $drop$, defined in the theory $List.thy$:

$$[\![s \uparrow_n]\!]_{Isab} = drop\ [\![n]\!]_{Isab}\ [\![s]\!]_{Isab}$$

For infinite streams we define Isabelle/HOL operator inf_drop:

```
constdefs
  inf_drop :: nat ⇒ (nat ⇒ 'a) ⇒ (nat ⇒ 'a)
  inf_drop i s ≡ λ j. s (i+j)
```

According to this definition, we get for an infinite stream s

$$[\![s \uparrow_n]\!]_{Isab} = inf\_drop \; [\![n]\!]_{Isab} \; [\![s]\!]_{Isab}$$

The syntax of the operators *drop* and *inf_drop* is the same for timed and untimed streams, but the semantics is different: in the timed case these operators remove the first k *time intervals*, where in the untimed case they remove the first k *elements* of the stream.

2.5.5. Timed Merge

We introduce the *timed merge operator* $\mathsf{merge}^{ti}(s,r)$, which concatenates the sequences of messages that are present on the channels (streams) s and r at the same time interval:

$$\mathsf{merge}^{ti} \in M^{\underline{\omega}} \times M^{\underline{\omega}} \to M^{\underline{\omega}}$$

Formally it is defined in FOCUS syntax as follows:

$$\forall \, t. \; \mathsf{ti}(\mathsf{merge}^{ti}(s,r),t) = \mathsf{ti}(s,t) ^\frown \mathsf{ti}(r,t) \tag{2.5}$$

For example, let

$$s_1 = \langle a_1, \sqrt{}, a_2, a_3, \sqrt{}, a_4, a_5, a_6, \sqrt{}, \sqrt{}, a_7, \sqrt{} \dots \rangle, \quad \text{and}$$
$$s_2 = \langle b_1, b_2, \sqrt{}, b_3, \sqrt{}, b_4, \sqrt{}, \sqrt{}, b_5, \sqrt{} \dots \rangle,$$

then

$$\mathsf{merge}^{ti}(s_1,s_2) = \langle a_1, b_1, b_2, \sqrt{}, a_2, a_3, b_3, \sqrt{}, a_4, a_5, a_6, b_4, \sqrt{}, \sqrt{}, a_7, b_5, \sqrt{} \dots \rangle$$

In Isabelle/HOL we represent this operator as follows:

✓ For finite streams we define the operator *fin_merge_ti*,
$[\![\mathsf{merge}^{ti}(s_1,s_2)]\!]_{Isab} \equiv fin\_merge\_ti \; [\![s_1]\!]_{Isab} \; [\![s_2]\!]_{Isab}$:

```
consts
  fin_merge_ti :: 'a fstream ⇒ 'a fstream ⇒ 'a fstream
primrec
  fin_merge_ti [] y = y
  fin_merge_ti (x#xs) y =
    ( case y of [] ⇒ (x#xs)
            | (z#zs) ⇒ (x@z) # (fin_merge_ti xs zs))
```

✓ For infinite streams we define the operator *inf_merge_ti*,
$[\![\mathsf{merge}^{ti}(s_1,s_2)]\!]_{Isab} \equiv inf\_merge\_ti \; [\![s_1]\!]_{Isab} \; [\![s_2]\!]_{Isab}$:

```
constdefs
  inf_merge_ti :: 'a istream ⇒ 'a istream ⇒ 'a istream
  inf_merge_ti x y
    ≡
    λ i. (x i)@(y i)
```

2.5.6. Concatenation of Time Intervals

The operator $\mathsf{ti}^k(s,n)$ denotes the sequence of messages that are present on the channel s at the time interval between ticks $n-1$ and $n+k$:

$$\mathsf{ti}^k(s, n) = \mathsf{ti}(s, n) \frown \ldots \frown \mathsf{ti}(s, n+k)$$

We define this operator formally in FOCUS syntax as follows:

$$\mathsf{ti}^k(s, n) \;\stackrel{\text{def}}{=}\; \begin{cases} \mathsf{ti}(s, n) & \text{if } k = 0 \\ \mathsf{ti}^{k-1}(s, n) \frown \mathsf{ti}(s, n+k) & \text{otherwise} \end{cases} \tag{2.6}$$

The Isabelle/HOL representation of the operator $\mathsf{ti}^k(s,n)$ for finite and infinite streams is identical modulo type of streams and is given below.

$$
\begin{aligned}
[\![\mathsf{ti}^0(s, n)]\!]_{Isab} \;&\equiv\; join\_ti \; [\![s]\!]_{Isab} \; [\![n]\!]_{Isab} \; [\![0]\!]_{Isab} \\
&\equiv\; [\![\mathsf{ti}(s, n)]\!]_{Isab}
\end{aligned}
$$

$$
\begin{aligned}
[\![\mathsf{ti}^{i+1}(s, n)]\!]_{Isab} \;&\equiv\; [\![\mathsf{ti}^i(s, n) \frown \mathsf{ti}(s, n+i+1)]\!]_{Isab} \\
&\equiv\; [\![\mathsf{ti}^i(s, n)]\!]_{Isab} @ [\![\mathsf{ti}(s, n+i+1)]\!]_{Isab}
\end{aligned}
$$

For finite streams:

$$
[\![\mathsf{ti}^k(s, n)]\!]_{Isab} \;\equiv\; fin\_join\_ti \; [\![s]\!]_{Isab} \; [\![n]\!]_{Isab} \; [\![k]\!]_{Isab},
$$

more precisely,

$$
\begin{aligned}
[\![\mathsf{ti}^0(s, n)]\!]_{Isab} \;&\equiv\; fin\_join\_ti \; [\![s]\!]_{Isab} \; [\![n]\!]_{Isab} \; [\![0]\!]_{Isab} \\
&\equiv\; nth \; [\![s]\!]_{Isab} \; [\![n]\!]_{Isab} \\
[\![\mathsf{ti}^{i+1}(s, n)]\!]_{Isab} \;&\equiv\; fin\_join\_ti \; [\![s]\!]_{Isab} \; [\![n]\!]_{Isab} \; [\![i+1]\!]_{Isab} \\
&\equiv\; [\![\mathsf{ti}^i(s, n)]\!]_{Isab} @ [\![\mathsf{ti}(s, n+i+1)]\!]_{Isab} \\
&\equiv\; fin\_join\_ti \; [\![s]\!]_{Isab} \; [\![n]\!]_{Isab} \; [\![i]\!]_{Isab} @ nth \; [\![s]\!]_{Isab} \; [\![n+i+1]\!]_{Isab}
\end{aligned}
$$

```
consts
  fin_join_ti ::'a fstream ⇒ nat ⇒ nat ⇒ 'a list
primrec
fin_join_ti_0:
  fin_join_ti s x 0 = nth s x
fin_join_ti_Suc:
  fin_join_ti s x (Suc i) = (fin_join_ti s x i) @ (nth s (x + (Suc i)))
```

For infinite streams:

$$\llbracket \mathsf{ti}^k(s, n) \rrbracket_{Isab} \quad \equiv \quad join\_ti \; \llbracket s \rrbracket_{Isab} \; \llbracket n \rrbracket_{Isab} \; \llbracket k \rrbracket_{Isab},$$

more precisely,

$$\llbracket \mathsf{ti}^0(s, n) \rrbracket_{Isab} \quad \equiv \quad join\_ti \; \llbracket s \rrbracket_{Isab} \; \llbracket n \rrbracket_{Isab} \; \llbracket 0 \rrbracket_{Isab}$$
$$\equiv \quad \llbracket s \rrbracket_{Isab} \; \llbracket n \rrbracket_{Isab}$$
$$\llbracket \mathsf{ti}^{i+1}(s, n) \rrbracket_{Isab} \quad \equiv \quad join\_ti \; \llbracket s \rrbracket_{Isab} \; \llbracket n \rrbracket_{Isab} \; \llbracket i + 1 \rrbracket_{Isab}$$
$$\equiv \quad \llbracket \mathsf{ti}^i(s, n) \rrbracket_{Isab} @ \llbracket \mathsf{ti}(s, n + i + 1) \rrbracket_{Isab}$$
$$\equiv \quad (join\_ti \; \llbracket s \rrbracket_{Isab} \; \llbracket n \rrbracket_{Isab} \; \llbracket i \rrbracket_{Isab}) @ (\llbracket s \rrbracket_{Isab} \; (\llbracket n + i + 1 \rrbracket_{Isab}))$$

consts
  $join\_ti ::'a \; istream \Rightarrow nat \Rightarrow nat \Rightarrow {}'a \; list$
primrec
$join\_ti\_0$:
  $join\_ti \; s \; x \; 0 = s \; x$
$join\_ti\_Suc$:
  $join\_ti \; s \; x \; (Suc \; i) = (join\_ti \; s \; x \; i) \; @ \; (s \; (x + (Suc \; i)))$

A number of properties of this operator (for the finite and the infinite cases) is proved in the theory *join_ti.thy* (see Appendix A.2). We present here one of these properties as Lemma 2.1.

Lemma 2.1:
If the operator $\mathsf{ti}^k(s,n)$ yields am empty stream $\langle\rangle$, every jth time interval, $n \leq j \leq n + k$, is empty:

$$\mathsf{ti}^k(s, n) = \langle\rangle \rightarrow \forall i \leq k : \mathsf{ti}(s, n + i) = \langle\rangle$$

The opposite also holds:

$$(\forall i \leq k : \mathsf{ti}(s, n + i) = \langle\rangle) \rightarrow \mathsf{ti}^k(s, n) = \langle\rangle$$

$\square$

2.5.7. Limited Number of Messages

The predicate $\mathsf{msg}_n(s)$ is **true** iff the stream s has at every time interval at most n messages. This predicate can be used both in Focus and in Janus.

$$\mathsf{msg} \in \mathbb{N} \times M^{\underline{\omega}} \rightarrow \mathbb{B}$$
$$\mathsf{msg}_n(s) \stackrel{\text{def}}{=} \forall t \in \mathbb{N}. \; \#\mathsf{ti}(s, t) \leq n \tag{2.7}$$

The Isabelle/HOL representation of this predicate has also two cases, for infinite and for finite streams:

constdefs
   msg :: nat ⇒ 'a istream ⇒ bool
   msg n s ≡ ∀ t. length (s t) ≤ n

consts
   fin_msg :: nat ⇒ 'a list list ⇒ bool
primrec
   fin_msg n [] = True
   fin_msg n (x#xs) = (((length x) ≤ n) ∧ (fin_msg n xs))

Lemma 2.2:

For any infinite timed stream s of type M the following relation holds: if the stream s has at every time interval at most one message, then every nonempty tth time interval, $\mathsf{ti}(s, t)$, consists exactly of one message, which can be defined as the first element of this time interval: $\mathsf{ft.ti}(s, t)$.

$$\mathsf{msg}_1(s) \Rightarrow$$
$$\forall\, t : \mathsf{ti}(s, t) \neq \langle\rangle \Rightarrow \exists\, a \in M : \mathsf{ti}(s, t) = \langle a \rangle \wedge a = \mathsf{ft.ti}(s, t) \tag{2.8}$$

Proof

By the definition of the predicate msg the stream s can have at every time interval at most one message, i.e. the message sequence representing a time interval can be either empty or it can contain exactly one message. The message sequence x representing a time interval t is nonempty. Thus, this sequence contains exactly one message, which is by the definition of the operator ft. the first element of this sequence. The proof of Lemma 2.2 in Isabelle/HOL is presented in Appendix A.1.

□

Additionally we have shown for every stream s, that if the predicate $\mathsf{ts}(s)$ holds, then the predicate $\mathsf{msg}_1(s)$ also must hold:

ts p ⟹ msg 1 p

 Remark: The relation 2.8 does not hold in the opposite direction:

$$\neg(\mathsf{ti}(s, t) = x \wedge x \neq \langle\rangle) \Leftarrow \exists\, a \in M : \mathsf{ti}(s, t) = \langle a \rangle \wedge a = \mathsf{ft}.x)$$

The message sequence $\mathsf{ti}(s, t) = x$ contains exactly one message a, which is a first element of some sequence x. From this we can conclude, that the sequence x is nonempty, but this does not imply that $\mathsf{ti}(s, t) = x$.

2.5.8. Stuttering Removal Operator for Timed Streams

As mentioned in Section 2.4.12, the stuttering removal operator $\propto$ operator is defined in Focus only for untimed streams. We define here an extended version of the stuttering removal operator for timed streams:

$$\propto^T \in M^{\underline{\omega}} \to M^{\underline{\omega}}$$

$\propto^T$ removes all duplications in every time interval. The Focus $\sqrt{}$ will be treated as a special element and not removed:

$$\forall t: \ \mathsf{ti}(\propto^T.s, t) \ = \ \propto(\mathsf{ti}(s, t))$$

In Isabelle/HOL we will have two definitions for this operator, for a finite and an infinite streams respectively. For the case of finite timed streams we define Isabelle/HOL function *finT_remdups*:

$$[\![\propto^T.s]\!]_{Isab} = finT\_remdups \ [\![s]\!]_{Isab}$$

and for the case of infinite timed streams we define Isabelle/HOL function *infT_remdups*

$$[\![\propto^T.s)]\!]_{Isab} = infT\_remdups \ [\![s]\!]_{Isab}$$

constdefs
  finT_remdups :: *'a fstream* $\Rightarrow$ *'a fstream*
  finT_remdups s $\equiv$ *map* $(\lambda \ s. \ remdups \ s) \ s$

  infT_remdups :: *'a istream* $\Rightarrow$ *'a istream*
  infT_remdups s $\equiv$ $(\lambda i.(\ remdups \ (s \ i)))$

2.5.9. Changing Time Granularity

In many cases it is useful to change time granularity of the specification (frequency of the streams, the time raster, see [Bro01], [Bro04]). For this reason we define new operators on timed streams to change the time granularity.

The operator $s \wedge_n$ refines the time granularity – it splits every time interval of the stream s into n time intervals in such a way that all messages from the original time interval belong to the first of the n intervals:

$$\wedge \in M^{\underline{\omega}} \times \mathbb{N} \to M^{\underline{\omega}}$$

$$\mathsf{ti}(s \wedge_n, t) \ \stackrel{\text{def}}{=} \ \begin{cases} \mathsf{ti}(s, t/n) & \mathsf{mod}(t, n) = 0 \\ \langle \rangle & \text{otherwise} \end{cases} \tag{2.9}$$

The operator $s \curlyvee_n$ makes the time granularity more coarse – it joins n time intervals of the stream s into a single time interval:

$$\curlyvee \in M^{\underline{\omega}} \times \mathbb{N}_+ \to M^{\underline{\omega}}$$

$$\text{ti}(s \, \curlyvee_n, t) \stackrel{\text{def}}{=} \text{ti}^{n-1}(s, n*t) \tag{2.10}$$

It is also possible to define another kinds of the operator $s \, \curlywedge_n$, e.g.

- ✓ all messages from the time interval of the original stream belong to the *last* of the n corresponding intervals,
- ✓ the messages from the time interval of the original stream are distributed to the n corresponding intervals.

Example 2.1:

Fig. 2.1 illustrates the duplication of the time raster of the stream x: $y = x \, \curlywedge_2$ (or $y \, \curlyvee_2 = x$), which is defined as simplification of Equations 2.9 and 2.10 using the equation $\text{mod}(t, 2) = 0 \Leftrightarrow \text{even}(t)$:

$$\text{ti}(x \, \curlywedge_2, t) \stackrel{\text{def}}{=} \begin{cases} \text{ti}(x, t/2) & \text{even}(t) \\ \langle\rangle & \text{otherwise} \end{cases}$$

$$\text{ti}(y \, \curlyvee_2, t) \stackrel{\text{def}}{=} \text{ti}(y, 2*t) \frown \text{ti}(y, 2*t+1)$$

□

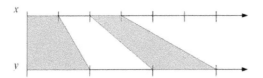

Figure 2.1.: Duplicated time raster

We define these operators in Isabelle/HOL for two kinds of timed streams, finite and infinite, in theories *fin_time_raster.thy* and *time_raster.thy* respectively:

- ✓ if s is a finite timed stream, we specify

  $[\![s \, \curlywedge_n]\!]_{Isab} \equiv \textit{fin\_split\_time} \, [\![s]\!]_{Isab} \, [\![n]\!]_{Isab}$

  and

  $[\![s \, \curlyvee_n]\!]_{Isab} \equiv \textit{fin\_join\_time} \, [\![s]\!]_{Isab} \, [\![n]\!]_{Isab}$

- ✓ if s is an infinite timed stream, we specify

  $[\![s \, \curlywedge_n]\!]_{Isab} \equiv \textit{split\_time} \, [\![s]\!]_{Isab} \, [\![n]\!]_{Isab}$

  and

  $[\![s \, \curlyvee_n]\!]_{Isab} \equiv \textit{join\_time} \, [\![s]\!]_{Isab} \, [\![n]\!]_{Isab}$

The Isabelle/HOL definitions of the functions *fin_split_time*, *fin_join_time*, *split_time* and *join_time* are presented in Appendix A.3.

Lemma 2.3:

For any infinite timed stream x and for any natural number $n > 0$ the following equation holds:

$$(x \downarrow_n) \curlyvee_n = x \tag{2.11}$$

$\square$

The proof of the lemma (for finite and infinite timed streams) in Isabelle/HOL is presented in Appendix A.3.

2.5.10. Deleting the First Time Interval

The operator ttl(s) denotes a timed stream which is obtained from the stream s deleting the first time interval:

$$\text{ttl} \in M^{\underline{\omega}} \to M^{\underline{\omega}}$$

We define this operator formally in Focus syntax as follows:

$$\forall\, t \in \mathbb{N} : \text{ti}(\text{ttl}(s), t) = \text{ti}(s, t+1) \tag{2.12}$$

The Isabelle/HOL semantics of this operator for the case of finite stream s is equal to the semantics Isabelle/HOL operator *tl* from the theory *List.thy*:

$$[\![\text{ttl}(s)]\!]_{Isab} \equiv tl\ [\![s]\!]_{Isab}$$

For the case of infinite timed stream s we define in Isabelle/HOL the corresponding operator *inf_tl*, $[\![\text{ttl}(s)]\!]_{Isab} \equiv inf\_tl\ [\![s]\!]_{Isab}$:

constdefs
   inf_tl :: $(nat \Rightarrow {}'a) \Rightarrow (nat \Rightarrow {}'a)$
   inf_tl $s \equiv (\lambda\ i.\ s\ (Suc\ i))$

2.5.11. First Nonempty Time Interval

The operators $\text{fti}^{\text{fin}}(s)$ and $\text{fti}^{\text{inf}}(s)$ denote the first nonempty time interval of a finite and an infinite timed stream respectively

$$\text{fti}^{\text{fin}} \in M^{\underline{*}} \to M^{*}$$
$$\text{fti}^{\text{inf}} \in M^{\underline{\infty}} \to M^{*}$$

To cover the exception that all the time intervals of a timed stream s are empty, we add to the formal Focus definition of this operator the rule "the operator returns an empty stream (an empty list), if the stream s consists of only empty time intervals":

$$
\begin{aligned}
\text{fti}^{\text{fin}}(s) = \quad &\text{if}\ \exists\, i \leq \text{Nti}(s) : \text{ti}(s, i) \neq \langle\rangle \\
&\text{then}\ \text{ti}(s, \min\{\{i \mid \text{ti}(s, i) \neq \langle\rangle\}\}) \\
&\text{else}\ \langle\rangle \quad \text{fi}
\end{aligned}
\tag{2.13}
$$

$$\mathsf{fti}^{\mathsf{inf}}(s) = \quad \text{if } \exists\, i \in \mathbb{N} : \mathsf{ti}(s, i) \neq \langle\rangle$$
$$\text{then } \mathsf{ti}(s, \min\{\{i \mid \mathsf{ti}(s, i) \neq \langle\rangle\}\}) \qquad (2.14)$$
$$\text{else } \langle\rangle \quad \text{fi}$$

The corresponding Isabelle/HOL definitions for the case of finite stream:

$[\![\mathsf{fti}^{\mathsf{fin}}(s)]\!]_{Isab}$

$\equiv [\![\text{if } \exists\, i \leq \mathsf{Nti}(s) : \mathsf{ti}(s, i) \neq \langle\rangle \text{ then } \mathsf{ti}(s, \min\{\{i \mid \mathsf{ti}(s, i) \neq \langle\rangle\}\}) \text{ else } \langle\rangle \text{ fi }]\!]_{Isab}$

$\equiv (if\, [\![\exists\, i \leq \mathsf{Nti}(s) : \mathsf{ti}(s, i) \neq \langle\rangle]\!]_{Isab}$

    $then\, [\![\mathsf{ti}(s, \min\{\{i \mid \mathsf{ti}(s, i) \neq \langle\rangle\}\})]\!]_{Isab}$

    $else\, [\![\langle\rangle]\!]_{Isab})$

$\equiv (if\, \exists\, i < [\![\mathsf{Nti}(s)]\!]_{Isab}.\, [\![\mathsf{ti}(s, i) \neq \langle\rangle]\!]_{Isab}$

    $then\, ti\, [\![s]\!]_{Isab}\, [\![\min\{\{i \mid \mathsf{ti}(s, i) \neq \langle\rangle\}\}]\!]_{Isab}$

    $else\, [])$

$\equiv (if\, \exists\, i < length\, [\![s]\!]_{Isab}.\, ti\, [\![s]\!]_{Isab}\, [\![i]\!]_{Isab} \neq [\![\langle\rangle]\!]_{Isab}$

    $then\, ti\, [\![s]\!]_{Isab}\, [\![\min\{\{i \mid \mathsf{ti}(s, i) \neq \langle\rangle\}\}]\!]_{Isab}$

    $else\, [])$

$\equiv (if\, \exists\, i < length\, s.\, ti\, s\, i \neq []$

    $then\, ti\, s\, (LEAST\, i.\, ti\, s\, i \neq [])$

    $else\, [])$

$\equiv fin\_find1nonemp\, s$

```
consts
  fin_find1nonemp :: 'a fstream ⇒ 'a list
primrec
  fin_find1nonemp [] = []
  fin_find1nonemp (x#xs) =
    ( if x = []
      then fin_find1nonemp xs
      else x )
```

The corresponding Isabelle/HOL definitions for the case of infinite stream:

$[\![\mathsf{fti}^{\mathsf{inf}}(s)]\!]_{Isab}$

$\equiv [\![\text{if } \exists\, i \in \mathbb{N} : \mathsf{ti}(s, i) \neq \langle\rangle \text{ then } \mathsf{ti}(s, \min\{\{i \mid \mathsf{ti}(s, i) \neq \langle\rangle\}\}) \text{ else } \langle\rangle \text{ fi }]\!]_{Isab}$

$\equiv (if\, [\![\exists\, i \in \mathbb{N} : \mathsf{ti}(s, i) \neq \langle\rangle]\!]_{Isab}$

    $then\, [\![\mathsf{ti}(s, \min\{\{i \mid \mathsf{ti}(s, i) \neq \langle\rangle\}\})]\!]_{Isab}$

    $else\, [\![\langle\rangle]\!]_{Isab})$

$$\equiv (\textit{if } \exists\,(i::nat).\ [\![\mathsf{ti}(s,i) \neq \langle\rangle]\!]_{Isab}$$
$$\qquad \textit{then } [\![s]\!]_{Isab}\ [\![\min\{\{i \mid \mathsf{ti}(s,i) \neq \langle\rangle\}\}]\!]_{Isab}$$
$$\qquad \textit{else } [])$$

$$\equiv (\textit{if } \exists\,(i::nat).\ s\ i \neq []$$
$$\qquad \textit{then } s\ (\textit{LEAST } i.\ s\ i \neq [])$$
$$\qquad \textit{else } [])$$

$$\equiv \textit{inf\_find1nonemp } s$$

constdefs
  inf_find1nonemp :: *'a istream* $\Rightarrow$ *'a list*
  inf_find1nonemp s
    $\equiv$
  $(\textit{ if } (\exists\ i.\ s\ i \neq [])$
    $\textit{then } s\ (\textit{LEAST } i.\ s\ i \neq [])$
    $\textit{else } [] \ )$

2.5.12. Index of the First Nonempty Time Interval

The operators $\mathsf{ind}_{\mathsf{fti}}^{\mathsf{fin}}(s)$ and $\mathsf{ind}_{\mathsf{fti}}^{\mathsf{inf}}(s)$ denote the index of the first nonempty time interval of a finite and an infinite streams respectively:

$$\mathsf{ind}_{\mathsf{fti}}^{\mathsf{fin}} \in M^{\underline{*}} \to \mathbb{N}$$
$$\mathsf{ind}_{\mathsf{fti}}^{\mathsf{inf}} \in M^{\infty} \to \mathbb{N}$$

To cover the exception that all the time intervals of a timed stream s are empty, we add to the formal Focus definition of this operator the rule "the operator $\mathsf{ind}_{\mathsf{fti}}^{\mathsf{fin}}(s)$ returns a number $n+1$, if the stream s consists of only n empty time intervals". The operator $\mathsf{ind}_{\mathsf{fti}}^{\mathsf{fin}}(s)$ returns 1, if the stream s is empty (contains no time intervals) – the result is the same as for the case "the first time interval is nonempty".

$$\mathsf{ind}_{\mathsf{fti}}^{\mathsf{fin}}(s) = \quad \textbf{if } \exists\,i \leq \mathsf{Nti}(s) : \mathsf{ti}(s,i) \neq \langle\rangle$$
$$\textbf{then } \min\{\{i \mid \mathsf{ti}(s,i) \neq \langle\rangle\}\} \qquad (2.15)$$
$$\textbf{else } 1 \quad \textbf{fi}$$

$$\mathsf{ind}_{\mathsf{fti}}^{\mathsf{inf}}(s) = \quad \textbf{if } \exists\,i \in \mathbb{N} : \mathsf{ti}(s,i) \neq \langle\rangle$$
$$\textbf{then } \min\{\{i \mid \mathsf{ti}(s,i) \neq \langle\rangle\}\} \qquad (2.16)$$
$$\textbf{else } 1 \quad \textbf{fi}$$

The corresponding Isabelle/HOL definitions are equal modulo syntax:

$$[\![\mathsf{ind}_{\mathsf{fti}}^{\mathsf{fin}}(s)]\!]_{Isab} = \textit{fin\_find1nonemp\_index } [\![s]\!]_{Isab}$$

consts
  fin_find1nonemp_index :: *'a fstream ⇒ nat*
primrec
 fin_find1nonemp_index [] = *0*
 fin_find1nonemp_index (*x#xs*) =
  (*if x* = []
   then Suc (*fin_find1nonemp_index xs*)
   else 0)

$$\llbracket \mathsf{ind}^{\mathsf{inf}}_{\mathsf{fti}}(s) \rrbracket_{Isab} = \mathit{inf\_find1nonemp\_index} \ \llbracket s \rrbracket_{Isab}$$

constdefs
 inf_find1nonemp_index :: *'a istream ⇒ nat*
 inf_find1nonemp_index s
  ≡
 (*if* (∃ *i. s i* ≠ [])
  then (*LEAST i. s i* ≠ [])
  else 0)

2.5.13. Last Nonempty Time Interval

We define an operator $\mathsf{last}^{ti}(s,t)$, which returns the last nonempty time interval of the timed stream s until the tth time interval. If until the tth time interval all intervals were empty, the empty message list is returned.

$$\mathsf{last}^{ti} \in M^{\underline{\omega}} \times \mathbb{N} \to M^{*}$$
$$\mathsf{last}^{ti}(s,0) = \mathsf{ti}(s,0)$$
$$\mathsf{last}^{ti}(s,t+1) = \mathsf{if} \ \mathsf{ti}(s,t+1) \neq \langle\rangle \ \mathsf{then} \ \mathsf{ti}(s,t+1) \ \mathsf{else} \ \mathsf{last}^{ti}(s,t) \ \mathsf{fi}$$

In Isabelle/HOL we will have two definitions for this operator, for finite and infinite streams respectively:

consts
 fin_last_ti :: (*'a list*) *list ⇒ nat ⇒ 'a list*
 inf_last_ti :: *'a istream ⇒ nat ⇒ 'a list*
primrec
 fin_last_ti s 0 = *hd s*
 fin_last_ti s (*Suc i*) =
  (*if s*!(*Suc i*) ≠ []
  then s!(*Suc i*)
  else fin_last_ti s i)

primrec
 inf_last_ti s 0 = *s 0*
 inf_last_ti s (*Suc i*) =
  (*if s* (*Suc i*) ≠ []
  then s (*Suc i*)
  else inf_last_ti s i)

2.6. tiTable

State transition tables are used in Focus to characterize relations between values in more readable way. The argumentation about streams in a classical Focus table treats *message-by-message*. We introduce here a new variant of the Focus [5] state transition tables – tiTable – to have a more clear representation of component assurances when one argues about a component in the time interval manner. The tiTable differs from the classical Focus table on the following points:

✓ The argumentation about streams treats on time intervals – the possible combinations of message sequences on component channels at the time interval t between ticks $t-1$ and t are treated.

✓ For the case, when some additional assumptions are needed an extra *Assumption* column can be used. It is needed only relations over the current values of the local (and state) variables and the current time intervals of the input streams.

Like in a standard Focus table, the names of streams (channels) and local or state variables in a tiTable must be equal to the names that are used in the corresponding specification.

A tiTable C looks in general as follows:

tiTable C: $\forall t \in \mathbb{N}$

	i_1	...	i_m	o_1	...	o_n	v_1'	...	v_p'	Assumption
1										
2										
...										
j	a_1	...	a_m	b_1	...	b_n	d_1	...	d_p	a
...										
N										

N denotes here the number of table lines (the number of possible combinations for the tth time interval), $i_1, \ldots, i_m$ denote the input channels, $o_1, \ldots, o_n$ denote the output channels and $v_1, \ldots, v_p$ denote the local variables of the component, for which the table C is defined. $a_1, \ldots, a_m$, $b_1, \ldots, b_n$, $d_1, \ldots, d_p$ denote here the corresponding values of the streams and local variables for the tiTable line j.

By v' we denote the value of the local variable v after the transition, i.e. the value at the time interval $t+1$.

In the case the weak causality is not enough and the *strong causality*, i.e. the argumentation about the values of output streams that are produced after some delay, is needed, e.g. "if a component C has on its input channel i at the time interval t a message sequence a, then at the time interval $t+k$, $1 \leq k$ it produces on the output channel o the message sequence b", the following notation for the tiTable-head must be used: if the column of an output stream

---

[5] A tiTable can be used in Janus in the same manner as in Focus, without any changes.

o_i is labeled by o_i^k, then the corresponding output will be produced with the delay k.

Arguing about time intervals we can ensure the causality property of a defined component (or system), specifying time interval values of output streams via input and local values from the previous (for the strong causality) or the same (for the weak causality) time intervals. Specifying a component by tiTable we ensure the weak causality, where specifying a component by tiTable *with delays* we ensure strong causality (if every output stream in the table is defined using some nonzero delay).

The tiTable C presents the most general case, where all the possible columns are presented. In most cases not all of the columns are used specifying a component. For the functional representation (see Equation 2.18) this means that the expression for an omitted column i_l will be also omitted, i.e. $\text{ti}(i_l, t) = a_l$ can be replaced by true.

The representation as a tiTable of a relation between input and output streams, and local variables can be schematically reformulated to the corresponding formula in a purely textual manner as shown below.

Definition: 2.1
The representation of a tiTable C in the functional form is defined as formula:

$$\bigwedge_{j=1}^{N} L_j^C \tag{2.17}$$

where the subformula L_j^C denotes the jth line of the table C and is specified as follows:

$$L_j^C \stackrel{\text{def}}{=} \begin{aligned} &(a \wedge \text{ti}(i_1, t) = a_1 \wedge \cdots \wedge \text{ti}(i_n, t) = a_n) \\ &\rightarrow \text{ti}(o_1, t) = b_1 \wedge \cdots \wedge \text{ti}(o_m, t) = b_m \ \wedge \ v_1' = d_1 \wedge \ldots v_p' = d_p \end{aligned} \tag{2.18}$$

□

Thus, we represent an externally defined (not inside the specification body, but as single element) tiTable C in plain text by a predicate *titable_C*.

---
titable_C

$i_1 \in I_1^{\infty}; \ldots; i_n \in I_m^{\infty};$
$v_1 \in V_1; \ldots; v_p \in V_p;$
$o_1 \in O_1^{\infty}; \ldots; o_n \in O_n^{\infty};$

L_1^C
$\ldots$
L_N^C

The corresponding predicate in Isabelle/HOL will have in general[6] p extra parameters to represent the local variables, because their values before and after

---
[6]In special case, where the values before computation only of k, $k < p$, local variables are used, we will need respectively k extra parameters.

computation will be treated separately. Thus, for a value of a local variable v at time t we will have $[\![v]\!]_{Isab} = v\_in\ t$ and $[\![v']\!]_{Isab} = v\_out\ t$.

> constdefs
> $titable\_C ::$
> $[\![I_1]\!]_{Isab}\ istream \Rightarrow \ldots \Rightarrow [\![I_n]\!]_{Isab}\ istream \Rightarrow$
> $(nat \Rightarrow [\![V_1]\!]_{Isab}) \Rightarrow \ldots \Rightarrow (nat \Rightarrow [\![V_p]\!]_{Isab}) \Rightarrow$
> $[\![O_1]\!]_{Isab}\ istream \Rightarrow \ldots \Rightarrow [\![O_n]\!]_{Isab}\ istream \Rightarrow$
> $(nat \Rightarrow [\![V_1]\!]_{Isab}) \Rightarrow \ldots \Rightarrow (nat \Rightarrow [\![V_p]\!]_{Isab}) \Rightarrow bool$
>
> $titable\_C\ i1 \ldots im \quad v1\_in \ldots vp\_in \quad o1 \ldots on \quad v1\_out \ldots vp\_out$
> $\equiv$
> $\forall\, t.\ [\![L_1^C]\!]_{Isab}\ \wedge\ \ldots\ \wedge\ [\![L_N^C]\!]_{Isab}$

Like in a standard FOCUS table, any auxiliary logical variable, which is needed in the tiTable, must be declared using the keyword univ [7]. If a tiTable is defined inside of the specification, the univ declaration of this variable must be written in the corresponding part of the FOCUS specification (see Example 2.2). For the case of externally defined tiTable the univ declaration must be written in brackets after the name of the tiTable, e.g. tiTable (univ $k \in \mathbb{N}$) *TimerT*1.

Using in a tiTable such an universally quantified variable, we can often omit the translation of input (stream) cells with such a variable, if this variable is not used in this line neither in the assumption nor to define the output (stream) cells (see Example 2.2 and Section 4.3.7 for more details).

⚠ **Remark:** From Lemma 2.2 follows: If the predicate $\mathsf{msg}_1(s)$ holds for some stream s, the notation x in a call of a tiTable that belongs to the column representing this stream s, together with the assumption $x \neq \langle\rangle$ is semantically equal to the notation $\langle a \rangle$ in this cell.

 Example 2.2:
Let us discuss a simple timer specification. The component *Timer* has one input channel s (of type $\mathbb{N}$) to set the timer and one output channel r (of type $\{timeout\}$) to signal the timeout. At every time interval it can receive at most one message of type $\mathbb{N}$. If the timer receives a message n, it waits n time intervals (counts down from n) and gives out the timeout signal *timeout*. If during countdown a new message comes, the timer will be reset.

The datatype $\{timeout\}$ can be specified in Isabelle/HOL as follows:

> $datatype\ timeoutType = timeout$

We specify the timer in FOCUS by tiTable:

---

[7]Any variable declared by the keyword univ is universally quantified within the scope of the table.

```
╔═ Timer ═══════════════════════════════════════ timed ═══
║   in      x : ℕ
║   out     y : {timeout}
╟──────────────────────────────────────────────────────────
║   local   l ∈ ℕ
║   univ    k ∈ ℕ
╟ ─ ─ ─ ─ ─ ─ ─ ─ ─ ─ ─ ─ ─ ─ ─ ─ ─ ─ ─ ─ ─ ─ ─ ─ ─ ─ ─ ─
║   init    l = 0
╟ ─ ─ ─ ─ ─ ─ ─ ─ ─ ─ ─ ─ ─ ─ ─ ─ ─ ─ ─ ─ ─ ─ ─ ─ ─ ─ ─ ─
║   asm
║     msg₁(x)
╟ ─ ─ ─ ─ ─ ─ ─ ─ ─ ─ ─ ─ ─ ─ ─ ─ ─ ─ ─ ─ ─ ─ ─ ─ ─ ─ ─ ─
║   gar
║     tiTable TimerT1 : ∀ t ∈ ℕ
```

$msg_1(x)$

gar — tiTable $TimerT1 : \forall\, t \in \mathbb{N}$

	x	y	l'	Assumption
1	$\langle k \rangle$	$\langle \rangle$	k	true
2	$\langle \rangle$	$\langle \rangle$	$l-1$	$l > 1$
3	$\langle \rangle$	$\langle timeout \rangle$	$l-1$	$l = 1$
4	$\langle \rangle$	$\langle \rangle$	0	$l = 0$

The formula below represents the semantics of the tiTable $TimerT1$.

$$
\begin{aligned}
\forall\, k, t : &\ (\mathsf{ti}(x,t) = \langle k \rangle \ \wedge\ \mathsf{true} \ \rightarrow\ \mathsf{ti}(y,t) = \langle \rangle \wedge l' = k) \\
\wedge\ &(\mathsf{ti}(x,t) = \langle \rangle \ \wedge\ l > 1 \ \rightarrow\ \mathsf{ti}(y,t) = \langle \rangle \wedge l' = l - 1) \\
\wedge\ &(\mathsf{ti}(x,t) = \langle \rangle \ \wedge\ l = 1 \rightarrow\ \mathsf{ti}(y,t) = \langle timeout \rangle \wedge l' = l - 1) \\
\wedge\ &(\mathsf{ti}(x,t) = \langle \rangle \ \wedge\ l = 0 \ \rightarrow\ \mathsf{ti}(y,t) = \langle \rangle \wedge l' = 0)
\end{aligned}
\tag{2.19}
$$

Because of the assumption $msg_1(x)$, we have that for every t

$$\mathsf{ti}(x,t) = \langle \rangle \vee \exists\, k \in \mathbb{N} : \mathsf{ti}(x,t) = \langle k \rangle$$

We can rewrite the formula 2.19 in the following way (the ∀-quantifier does not be needed here, because the variable k does not to be used any more)

$$
\begin{aligned}
\forall\, t : &\ (\mathsf{ti}(x,t) \neq \langle \rangle \ \rightarrow\ \mathsf{ti}(y,t) = \langle \rangle \wedge l' = \mathsf{ft}.\mathsf{ti}(x,t)) \\
\wedge\ &(\mathsf{ti}(x,t) = \langle \rangle \ \wedge\ l > 1 \ \rightarrow\ \mathsf{ti}(y,t) = \langle \rangle \wedge l' = l - 1) \\
\wedge\ &(\mathsf{ti}(x,t) = \langle \rangle \ \wedge\ l = 1 \rightarrow\ \mathsf{ti}(y,t) = \langle timeout \rangle \wedge l' = l - 1) \\
\wedge\ &(\mathsf{ti}(x,t) = \langle \rangle \ \wedge\ l = 0 \ \rightarrow\ \mathsf{ti}(y,t) = \langle \rangle \wedge l' = 0)
\end{aligned}
\tag{2.20}
$$

This logic formula can be also represented as an encapsulated "if_then_else" expression (some argue that such kind of representation is easier to comprehend as a logic formula like in the definition of the tiTable), because the tiTable $TimerT1$ covers all possible combinations of the left implication parts of subformulas (so, all "else"-parts of expressions are presented):

$$
\begin{aligned}
&\forall\, t : \\
&\text{if ti}(x, t) \neq \langle\rangle \\
&\text{then ti}(y, t) = \langle\rangle \wedge l' = \text{ft.ti}(x, t) \\
&\text{else if } l > 1 \\
&\quad\text{then ti}(y, t) = \langle\rangle \wedge l' = l - 1 \\
&\quad\text{else if } l = 1 \\
&\qquad\text{then ti}(y, t) = \langle timeout\rangle \wedge l' = l - 1 \\
&\qquad\text{else ti}(y, t) = \langle\rangle \wedge l' = 0 \\
&\qquad\text{fi} \\
&\quad\text{fi} \\
&\text{fi}
\end{aligned}
\tag{2.21}
$$

The semantically equal representation will be also the following one

$$
\begin{aligned}
&\forall\, t \in \mathbb{N} : \\
&l' = (\text{if ti}(s, t) = \langle\rangle \text{ then } l \text{ else ft.ti}(s, t) \text{ fi }) \\
&\text{ti}(r, t) = (\text{if } l = 0 \text{ then } \langle timeout\rangle \text{ else } \langle\rangle \text{ fi })
\end{aligned}
\tag{2.22}
$$

$\square$

In the general case, we can translate a tiTable to the plain text representation as a FOCUS predicate.

More examples of tiTables are given in Sections 4.1 and 4.3.

⚠ **Remark:** The tabular representation does not equal in general to an encapsulated "if_then_else" expression: the "if_then_else" expression assumes that the "else"-part is always defined, what is not the case for the tabular representation.

2.7. Encapsulated States

An encapsulated state in FOCUS is a tuple of attributes of arbitrary types. Encapsulated states are decomposed in FOCUS into three kinds of substates:

✓ *local states* that store some computed information (internal data memory of the component);

✓ *control states* that record the flow of control;

✓ *oracles* that capture nondeterminism.

The states are introduced in FOCUS by declaration state variables using the keyword local, and oracles are introduced using the keyword orac.

We define a FOCUS specification S with m input channels $i_1, \ldots, i_m$ of types $I_1, \ldots, I_m$, k output channels $r_1, \ldots, r_k$ of types $R_1, \ldots, R_k$, and with n state variables $st_1, \ldots, st_n$ of types $M_1, \ldots, M_n$ in Isabelle/HOL in the following way. First of all we define a predicate S_L adding $2n$ extra parameters to the signature of S, $stIn_1, \ldots, stIn_n$ and $stOut_1, \ldots, stOut_n$, where the jth element of the stream $stIn_i$ represents the value of the variable st_i before the j computation, and the jth element of the stream $stOut_i$ represents the value of the variable st_i after the jth computation.

These extra parameters can be seen as untimed infinite streams $stIn_i$ and $stOut_i$ of type M_i, $1 \leq i \leq n$, for which the following equations hold:

$$stIn_i = \langle InitV_i \rangle \frown stOut_i$$

where $InitV_i$ represents the initial values of the state variable st_i. More precisely, these streams are *time-synchronous*[8] ones – they have some message in every time interval. In the special case, where the type M represents a list of some type T, $M = T^*$, the message of these streams that can be equal to $\langle \rangle$ at some time intervals.

After that we define a predicate S in the following way:

constdefs
$S ::\ [\![I_1]\!]_{Isab}\ istream \Rightarrow \ldots [\![I_m]\!]_{Isab}\ istream \Rightarrow$
$\quad (nat \Rightarrow [\![M_1]\!]_{Isab}) \Rightarrow \ldots (nat \Rightarrow [\![M_n]\!]_{Isab}) \Rightarrow$
$\quad [\![R_1]\!]_{Isab}\ istream \Rightarrow \ldots [\![R_k]\!]_{Isab}\ istream \Rightarrow$
$\quad (nat \Rightarrow [\![M_1]\!]_{Isab}) \Rightarrow \ldots (nat \Rightarrow [\![M_n]\!]_{Isab}) \Rightarrow$
$\quad bool$
$S\ i1 \ldots im\ r1 \ldots rk\ \equiv$
$\quad (\exists\ l1 \ldots ln.$
$\quad\quad S\_L\ i1 \ldots in$
$\quad\quad (fin\_inf\_append\ [InitV1]\ l1) \ldots$
$\quad\quad (fin\_inf\_append\ [InitVn]\ ln)\ r1 \ldots rk\ l1 \ldots ln)$

⚠ **Remark:** A specification of an initial value (of a local or a state variable) has in some cases no influence on the computations and can be omitted in FOCUS. This kind of specification must be refined by fixing initial values of all local state variable to make the translation to Isabelle/HOL possible.

Example 2.3:
In the specification of ther timer (see Example 2.2) one local variable is used. Thus, to represent the specification *Timer* in Isabelle/HOL we need to add two extra parameters of type $\mathbb{N}$ to the corresponding predicate.

---

[8]In the original FOCUS meaning.

constdefs
Timer_L ::
  "nat istream $\Rightarrow$ (nat $\Rightarrow$ nat) $\Rightarrow$
  timeoutType istream $\Rightarrow$ (nat $\Rightarrow$ nat) $\Rightarrow$ bool"
"Timer x lIn y lOut
  $\equiv$
(msg (Suc 0) x)
  $\longrightarrow$
(lOut t = (if (x t = []) then (lIn t) else hd (x t)) $\wedge$
y t = (if (lIn t) = 0 then [timeout] else []))"

Now we can define the semantics of the specification *Timer*:

$$[\![(y) := Timer(x)]\!]_{Isab}$$
$$\equiv$$

constdefs
Timer ::
  nat istream $\Rightarrow$ timeoutType istream $\Rightarrow$ bool
Timer x y
  $\equiv$
($\exists$ l. Timer_L x (fin_inf_append [InitVn] l) y l)

                                                             $\square$

More examples are given in Sections 4.1 and 4.3.

Oracles differs from the local and control states in that they are independent of the information received trough the input channels. Thus, oracles are some ∃-quantified variables used in the specification (in contrast to the variables introduced by the keyword **univ** – the last ones are ∀-quantified in the specification). Therefore, we apply to the oracles the standard translation rules (see Sections 2.14 and 3.5).

2.8. Timed State Transition Diagrams

In this section we introduce *timed state transition diagrams* (TSTDs), which are visualizations of timed state transition tables (**tiTables**). The main difference between the proposed TSTDs and the standard state transition diagrams (STDs, see [GKRB96, KPR97]) is similar to the difference between a standard FOCUS table and a **tiTable** (see also Section 2.6) – specification of transitions between system states is based on time intervals. A node of a TSTD represents a control state. A transition of a TSTD is described by a row in the corresponding **tiTable**.

In comparison to the timed transition diagrams, introduced by T.A. Henzinger, Z. Manna and A. Pnueli (see [HMP91]), where for any transition a delay interval $[l, u]$ is specified ($l \in \mathbb{N}$ denotes a minimal delay, $u \in \mathbb{N} \cup \{\infty\}$ denotes a maximal delay), we are argue here about concrete delays, i.e. $l = u$. To specify the case, where some messages occurs in a stream s during some time interval $[l, u]$, where $l < u$, we can use the operator for concatenation of time intervals: $\mathsf{ti}^{u-l}(s, l)$ (see Section 2.5.6).

If some rows of a tiTable differ only in the assumption cells, they can be "compressed" in one row, s.t. the assumptions are composed by logical disjunction operator $\lor$. Thus, one tiTable line can describe a number of TSTD transitions. In some cases this property of a tiTable results more compact representation then by a TSTD. On the other hand represents a TSTD also an information, that is not contained in a tiTable – which of the control states is an initial one (for the case of specification by a tiTable, this must be defined by the keyword init in a FOCUS specification).

We suggest the following notation for TSTDs:

✓ The argumentation is over time intervals, the "current" time interval number is t, $t \in \mathbb{N}$.

✓ For any stream x, its tth time interval, $\mathsf{ti}(x, t)$, will be denoted on the TSTD labels by x. To denote $\mathsf{ti}(x, t+k)$ the abbreviation x^k can be used. The same notation is used in tiTables.

✓ For any *input* stream y (y belongs to the list of input channels used in the TSTD): if an expression of the form $\mathsf{ti}(y, t) = SomeTimeInterval$ is omitted, the value of the tth time interval of the stream y can be arbitrary.

✓ For any *output* stream z (z belongs to the list of output channels used in the TSTD) all expression of the form $\mathsf{ti}(z, t) = \langle \rangle$ are omitted.

✓ For any local (or state) variable l all expression of the form $l' = l$ are omitted.

Example 2.4:

The component *Kbit* has one input and one output channel, both of type $\mathbb{B}$it. The component stays in its initial state, $S0$, until the first nonempty time interval of the input stream. If it receives some messages in the initial state, it outputs in the same time interval the message 1 and goes to the next state, $S1$, where it stays also until the first nonempty time interval of the input stream. If it receives some messages in the state $S1$, it outputs them in the same time interval and goes to the state $S2$. The component *Kbit* stays in the state $S2$ until first empty time interval of the input stream and forwards all input messages to the output stream without any delay. If in some time interval no messages are received, it outputs message 0 and goes in the initial state $S0$.

We specify this component firstly by a timed STD and after that we present the corresponding tiTable.

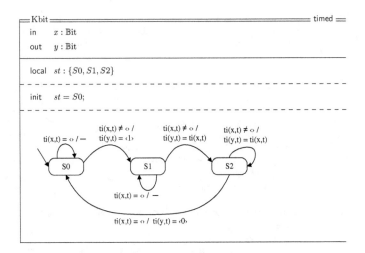

tiTable KbitT (univ r : Bit*): $\forall t \in \mathbb{N}$

	x	y	st'	Assumption
1	$\langle\rangle$	$\langle\rangle$	$S0$	$st = S0$
2	r	$\langle 1\rangle$	$S1$	$st = S0 \wedge r \neq \langle\rangle$
3	$\langle\rangle$	$\langle\rangle$	$S1$	$st = S1$
4	r	r	$S2$	$st = S1 \wedge r \neq \langle\rangle$
5	r	r	$S2$	$st = S2 \wedge r \neq \langle\rangle$
6	$\langle\rangle$	$\langle 0\rangle$	$S0$	$st = S2$

It easy to see, that the 4th and the 5th rows of the **tiTable** KbitT differ only by the assumption cells. Thus, we can specify the semantically equal **tiTable** KbitT_c, which is a "compressed" version of KbitT.

tiTable KbitT_c (univ r : Bit*): $\forall t \in \mathbb{N}$

	x	y	st'	Assumption
1	$\langle\rangle$	$\langle\rangle$	$S0$	$st = S0$
2	r	$\langle 1\rangle$	$S1$	$st = S0 \wedge r \neq \langle\rangle$
3	$\langle\rangle$	$\langle\rangle$	$S1$	$st = S1$
4	r	r	$S2$	$(st = S1 \vee st = S2) \wedge r \neq \langle\rangle$
5	$\langle\rangle$	$\langle 0\rangle$	$S0$	$st = S2$

□

A more extensive example of a TSTD is given in the Automotive-Gateway case study (see Section 4.3.7).

2.9. Mutually Recursive Functions

Specification of mutually recursive functions in Isabelle/HOL and especially the argumentation about them is extensive: first of all the corresponding mutually recursive datatypes (which will be basis for such a recursion) must be defined and only after that we can define the mutually recursive functions (see [NPW02] for details).

Mutually recursive functions can be used in a FOCUS specification to represent some local (internal) states of a component (see [BS01]). Thus, to solve the problem with translation of specified mutually recursive functions to Isabelle/HOL, we can simply use in FOCUS specification local variables (see Section 2.7) instead of mutual recursion.

For example, let two mutually recursive functions are used in some (existing) FOCUS specification of a component: one of them is chosen for the next step (e.g. to process the rest of some stream) according to the current values of component streams or states. We specify these functions as a joint function with extra parameter to deal with the current state (corresponds to the values of component streams or states) and add to the specification a new local variable to encode which part of the joint function must be used in the next step. This variable must be set initially to the value representing the corresponding part of joint function. An example of such a transformation is given in Section 4.1.6.

2.10. Time-Synchronous Streams

As mentioned in Section 2.2, time-synchronous streams are a special kind of streams, where each message represents one time interval. In FOCUS these streams are represented as untimed streams given a particular timed interpretation, i.e. their semantics is timed, but the syntax is equal to the syntax of untimed streams – a time-synchronous stream does not contain any $\sqrt{}$.

Combination of timed and time-synchronous streams leads to problems with type checking: the FOCUS time-synchronous streams cannot be combined in a composite FOCUS specification with any other kind of streams, neither with untimed streams, because of different semantics, nor with timed streams, because of different syntax. To have a possibility to use together a specification with time-synchronous streams and a specification with timed stream, we suggest to model a time-synchronous stream as a *timed* stream for which a predicate *ts* holds.

Thus, if we have a completed FOCUS specification with frame label "time-synchronous", the following changes must be done:

1. The specification frame label *time-synchronous* must be replaced by the label *timed*.

2. The assumption that *all* input streams are time-synchronous, as well as the guarantee that *all* output streams are also time-synchronous must be added.

3. The argumentation over elements of the stream must be replaced by the argumentation over the first (and unique) element of the corresponding time stream:

   $$s.t \longmapsto \mathsf{ft.ti}(s, t)$$

   If this argumentation was over the set $\mathbb{N}_+$ (the elements in a stream are counted from 1), we will argue now over the set $\mathbb{N}$, because the time intervals are counted from 0 (see Section 2.5).

For the case of a FOCUS specification with frame label "timed", where some of its input and/or output streams are time-synchronous, the changes are more minor:

1. The assumption that the *corresponding* input streams are time-synchronous, as well as the guarantee that the *corresponding* output streams are also time-synchronous must be added.

2. The argumentation over elements of the *corresponding* stream must be replaced to the argumentation over the first (and unique) element of the corresponding time stream:

   $$s.t \longmapsto \mathsf{ft.ti}(s, t)$$

   If this argumentation was over the set $\mathbb{N}_+$, we will argue now over the set $\mathbb{N}$.

2.11. Isabelle/HOL Semantics of Elementary and Composite FOCUS Specifications

We can represent the FOCUS styles of elementary and composite specifications in Isabelle/HOL in the following way:

✓ The *equational* and the *relational styles* are similar to the style that is used in Isabelle. Thus, we represent a FOCUS component that is described in equational or relational style in Isabelle/HOL as a predicate.

✓ We represent a FOCUS component that is specified in *A/G style* by implication: the assumption predicate implies the guarantee predicate.

✓ A FOCUS specification in *graphical style* can be schematically translated into a FOCUS specification of equational or relational style.

The FOCUS semantics of an assumption part as well as the semantics of a guarantee part of a FOCUS specification in A/G style is a predicate representing relations over streams. The semantics of the *Body*-part of an FOCUS specification in A/G style is a predicate that is build as an implication $A \rightarrow G$, where A denotes the assumption predicate and G denotes the guarantee predicate.

⚠ **Remark:** The A/G style is a most general one and we suggest to use this style in the most cases. The only exception is the pure system architecture specification, which serves only to show in a readable way how the subcomponents are connected. If for some component we have not any assumption, we can also fill the assumption part with **true**. In such a way we can partially[9] solve the problem with forgotten assumptions (see Section 3.5).

2.11.1. Auxiliary Datatypes, Functions and Predicates

We will use the expression *specification group* to denote the set of

✓ specifications of the system on all refinement steps (both requirements and architecture specifications), and

✓ specifications of all subcomponents (or, more precisely, all subcomponent trees) on all refinement steps.

To represent the Isabelle/HOL semantics of the lists (sets) of input, output and local channel identifiers of a FOCUS specification S, we need to define first of all the following datatypes (using Isabelle/HOL keyword ***datatype***, see Section 1.2)

✓ *chanID* – the type of all needed channel identifiers $Id_1, \ldots, Id_N$:

---

[9] Here the "human factor" is taken into account: having the assumption-part in a specification, one needs to reflect how this part must be filled out. Thus, the probability to forget the necessary assumptions and to write **true** to fill out this part instead of the assumptions is much less in comparison to the possibility to use any kind ("Assumption/Guarantee" or only "Guarantee") of specification.

datatype *chanID* =
  $ch\_Id_1 \mid \cdots \mid ch\_Id_N$

✓ *specID* – the type of all needed specification names $S_1, \ldots, S_M$ (specification identifiers):

datatype *specID* =
  $sS_1 \mid \cdots \mid sS_M$

These types must be defined for the specification group as whole in the theory *SpecificationGroupName_types.thy* (see Sections 2.11.5). Then we specify the sets of input, output and local identifiers of the specification S as functions ins[10], *out* and *loc*, which returns for a given specification name a set of corresponding channel identifiers – input, output and local respectively. For an elementary specification the function *loc* returns an empty set.

After that we introduce predicates *inStream*, *outStream* and *locStream* over specification names and channel identifiers. The predicate *inStream S x* (*outStream S x*, *locStream S x*) is *True*, if the set of channel identifiers x corresponds to the specification name S.

We also specify a function *subcomponents* which returns for a given component a set of its subcomponents. For an elementary specification the function *subcomponents* returns an empty set.

2.11.2. Elementary Specification

The definition of the Focus semantics of an elementary timed specification S (see Definition 1.1) with the input channels $i_1, \ldots, i_n$ and the output channels $o_1, \ldots, o_m$ (in the case of parameterized specification: also with parameters $p_1, \ldots, p_k$ for some $k \in \mathbb{N}$):

```
══S (const  p₁ ∈ P₁; …;  pₖ ∈ Pₖ)═══════════════════ timed ══
  in     i₁ : I₁;  …;   iₙ : Iₙ
  out    o₁ : O₁;  …;   oₘ : Oₘ
  ─────────────────────────────────────────────────────────
  B_S
```

can be represented in Isabelle/HOL as the following conjunction

$$[\![i_S]\!]_{Isab} \ \wedge \ [\![o_S]\!]_{Isab} \ \wedge \ [\![B_S]\!]_{Isab} \tag{2.23}$$

where

----

[10]The name *in* is a predefined Isabelle/HOL expression and cannot be used here as a function name.

✓ $[\![B_S]\!]_{Isab}$ is an Isabelle/HOL predicate representing the Isabelle/HOL semantics of the body B_S of the FOCUS specification S.

$[\![B_S]\!]_{Isab}$ describes here the relation (in the case of parameterized specification: with k extra parameters) between the input and output streams and is equal modulo syntax to B_S;

✓ $[\![i_S]\!]_{Isab}$ is a predicate representing the Isabelle/HOL semantics of the set of input identifiers i_S of the FOCUS specification S:

$$[\![i_S]\!]_{Isab} \stackrel{\text{def}}{=} \text{inStream S (ins S)} \qquad (2.24)$$

✓ $[\![o_S]\!]_{Isab}$ is a predicate representing the Isabelle/HOL semantics of the set of output identifiers o_S of the FOCUS specification S:

$$[\![o_S]\!]_{Isab} \stackrel{\text{def}}{=} \text{outStream S (out S)} \qquad (2.25)$$

The body B_S of an elementary[11] FOCUS specification S (except the A/G specification) represented by a number of logic formulas – each formula represents some logic property – looks as follows:

P_1

P_2

. . .

P_n

The line breaks denote in FOCUS conjunction. Thus, the representation of B_S in Isabelle/HOL will be conjunction of the Isabelle/HOL representations of the properties:

$$[\![B_S]\!]_{Isab} = \bigwedge_{i=1}^{n} [\![P_i]\!]_{Isab} \qquad (2.26)$$

In the case of the A/G specification the body B_S consists of two parts – the assumption part B_S^A and the guarantee part B_S^G. The semantics of such a specification in FOCUS is logical implication

$$[\![B_S^{AG}]\!]_{Isab} = [\![B_S^A]\!]_{Isab} \longrightarrow [\![B_S^G]\!]_{Isab} \qquad (2.27)$$

where for B_S^A and B_S^G hold the presented above rules for a non-A/G specification.

The suggested order of the parameters in the relation is the following one: number of channels in the sheaf, input streams, specification parameters, output streams.

---

[11]The representation of an elementary *parameterized* FOCUS specification is analog.

For the proofs of the properties of the specification S we need only the predicate *body* that represents $[\![B_S]\!]_{Isab}$. Therefore, only this part will be denoted later as Isabelle/HOL semantic of the specification:

$$[\![S]\!]_{Isab} \stackrel{\text{def}}{=} [\![B_S]\!]_{Isab} \tag{2.28}$$

More precisely, the representation of the FOCUS specification S in Isabelle/HOL will have the following form:

constdefs
$$S :: [\![I_1]\!]_{Isab} \Rightarrow \ldots \Rightarrow [\![I_n]\!]_{Isab} \Rightarrow$$
$$[\![P_1]\!]_{Isab} \Rightarrow \ldots \Rightarrow [\![P_k]\!]_{Isab} \Rightarrow$$
$$[\![O_1]\!]_{Isab} \Rightarrow \ldots \Rightarrow [\![O_m]\!]_{Isab} \Rightarrow Bool$$
$$S \; i_1 \ldots i_n p_1 \ldots p_k o_1 \ldots o_m$$
$$\equiv$$
$$[\![B_S]\!]_{Isab}$$

Correctness of the relations between the sets of input and output channels, $[\![i_S]\!]_{Isab}$ and $[\![o_S]\!]_{Isab}$, as well as $[\![l_S]\!]_{Isab}$, will be shown separately in the Isabelle/HOL theory *SpecificationGroupName_inout.thy* (see Sections 2.11.5).

We define the Isabelle/HOL semantics of an *elementary parameterized* timed specification (see Definition 1.2 for the definition for FOCUS representation) with the input channels $i_1, \ldots, i_n$ and the output channels $o_1, \ldots, o_m$ in Isabelle, and with parameters $p_1, \ldots, p_k$ analogous, but now the Isabelle/HOL predicates in the relation $[\![B_S]\!]_{Isab}$ correspondingly have k extra parameters.

For a *time-synchronous* elementary specification with the input channels i_1, $\ldots, i_n$ and the output channels $o_1, \ldots, o_m$ we need additional constraints to define that the input and output streams are time-synchronous ones:

$$[\![S]\!]_{Isab} \stackrel{\text{def}}{=} \bigwedge_1^n \mathsf{ts}(i_j) \; \wedge \; \bigwedge_1^m \mathsf{ts}(o_j) \; \wedge \; [\![B_S]\!]_{Isab} \tag{2.29}$$

⚠ **Remark:** For the specifications with sheaves of channels extra additional constraints are needed - we need to make sure that the sheaves are nonempty, i.e. that every sheaf contains least one channel (see Sections 2.13 for more details).

The specification frames influence not only the semantics, but also syntax of the specification. The frame *untimed* (*timed, time-synchronous*) means that all input, output and local streams of the specified component must be untimed (timed, time-synchronous). The frame *untimed* implies, that all operations on the timed (and also time-synchronous) streams must be excluded. The frame *timed* does not exclude that some of these streams behave like time-synchronous ones, i.e. contain at each time interval exactly one message.

⚠ **Remark:** For any FOCUS specification S the sets i_S and o_S must be disjoint:

$$i_S \cap o_S = \varnothing \qquad (2.30)$$

2.11.3. Composite Specification

We define the Isabelle/HOL semantics of a *composite* specification S in Isabelle/HOL accordingly to the corresponding definition in FOCUS (see Definition 1.3):

$$[\![S]\!]_{Isab} \stackrel{\text{def}}{=} \exists\, l_S \in L_S : \bigwedge_{j=1}^{n} [\![S_j]\!]_{Isab} \qquad (2.31)$$

where l_S denotes a set of *local streams* and L_S denotes their corresponding types, $[\![S_j]\!]_{Isab}$ denotes the Isabelle/HOL semantics of the FOCUS specification S_j, $1 \le j \le n$, which is a specification of subcomponent of S.

⚠ **Remark:** For any composite FOCUS specification S the sets i_S, o_S and l_S must be pairwise disjoint, i.e. additionally to Equation 2.30 also the following equations must hold:

$$\begin{aligned} i_S \cap l_S &= \varnothing \\ l_S \cap o_S &= \varnothing \end{aligned} \qquad (2.32)$$

Equation 2.32 trivially holds for any elementary specification, because for any elementary specification S the set l_S is empty. Thus, Equations 2.30 and 2.32 build together the common property of correct relations between the sets of input, output and local channels that we specify in Isabelle/HOL as predicate *correctInOutLoc* over *specID*. We need to prove that this predicate holds for every specification in the system.

The sets i_S and o_S of input and output channel identifiers of a composite specification S consist of all sets of input and output channel identifiers of composing specifications $S_1, \ldots, S_n$ excluding the channels which are used for the local communication:

$$i_S \stackrel{\text{def}}{=} \bigcup_{j=1}^{n} (i_{S_j} \in I_S^{\infty}) \setminus l_S \qquad (2.33)$$

$$o_S \stackrel{\text{def}}{=} \bigcup_{j=1}^{n} (o_{S_j} \in O_{S_j}^{\infty}) \setminus l_S \qquad (2.34)$$

For the specification of the system S that is composed from the specifications $S_1, \ldots, S_n$ the following properties must hold:

✓ For the set of input streams of the system S:
Equation 2.33 holds. No input stream i can be an output stream of any subcomponent.

$$i_S = \bigcup_{j=1}^{n} (i_{S_j} \in I_{S}^{\infty}) \setminus l_S \ \wedge \ i_S \cap \bigcup_{j=1}^{n} o_{S_j} = \varnothing \qquad (2.35)$$

We define this property in Isabelle/HOL by the predicate on the specification identifiers, *CorrectCompositionIn* (see Section 2.11.4).

✓ For the set of output streams of the system S:
Equation 2.34 holds. No output stream i of the system S can be an input stream of any subcomponent.

$$o_S = \bigcup_{j=1}^{n} (o_{S_j} \in O_{S_j}^{\infty}) \setminus l_S \ \wedge \ o_S \cap \bigcup_{j=1}^{n} i_{S_j} = \varnothing \qquad (2.36)$$

The predicate *CorrectCompositionOut* on the specifications identifiers (see Section 2.11.4) presents this property in Isabelle/HOL.

✓ Every local stream l of the system S must be both an input stream of some subcomponent S_{j_1}, $1 \leq j_1 \leq n$, and an output stream of some subcomponent S_{j_2}, $1 \leq j_2 \leq n$ ($j_1 \neq j_2$):

$$l_S = \bigcup_{j=1}^{n} i_{S_j} \cap \bigcup_{j=1}^{n} o_{S_j} \qquad (2.37)$$

This property is specified Isabelle/HOL by the predicate on the specifications identifiers, *CorrectCompositionLoc* (see Section 2.11.4).

We also define a predicate *correctComposition* (see Section 2.11.4) that describes the following property: if a component S has no subcomponents, *subcomponents spec_S = []*, the set of local channels of this component, *loc spec_S*, must be empty.

The fulfillment of the properties presented above must be proved for every composite specification (see Sections 2.11.5).

 Remark: Equation 2.32 means amongst others, that the situation where some output stream of a component S is at the same time an input stream for some subcomponent of S is excluded. This restriction allows us to have clear separation of the global (input and output) and local streams (channels), which is very important to argue about system interfaces in the precise, formal, way.

Example 2.5:
Assuming a system that consists of three components: A, B and C and has two inputs (i_1 and i_2) and two outputs (o_1 and o_2). The representation S (see

Figure 2.2) is not correct, because the stream o_2, the output of the component C, is an input of the component B and at the same time an output of the system. The representation S', where the extra-output l of the component C is defined in the specification of C to be equal to the stream o_2, is correct. □

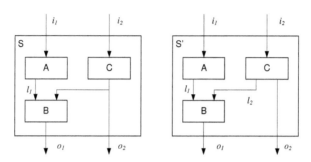

Figure 2.2.: Wrong and correct specification of subcomponents

2.11.4. Relations between Sets of Channels

In this section we discuss the Isabelle/HOL specification of relations between sets of channels for the cases without specification replication and sheaves of channels. The Isabelle/HOL specification of relation between sets of channels and sheaves of channels as well as specification replication will be discussed in Section 2.13.3.

The Isabelle/HOL theory *inout.thy* uses the (user-defined) theory *SpecificationGroupName_types.thy*, which contains all user-defined datatypes (including *specID* and *chanID*) used in the the specification group. This theory is a general one and can be used without any changes (except the name of the theory "SpecificationGroupName" that can be seen as parameter). We define here the signatures of the functions *subcomponents, ins, loc,* and *out,* as well as all predicates that specify the correctness properties of the relations between the sets of channels (see Sections 2.11.2 and 2.11.3): the predicates *correctInOutLoc* and *correctComposition* must hold for all components of the specification group, where the predicates *correctCompositionIn, correctCompositionOut* and *correctCompositionLoc* must hold only for composite components.

The functions *subcomponents, ins, loc,* and *out* will be defined in the theory *SpecificationGroupName_inout.thy* (that must be based on the theory *inout.thy*) by primitive recursion on the specification identifiers (see Section 2.11.5 for an example).

theory *inout = Main + SteamBoiler_types*:

consts
  subcomponents :: *specID ⇒ specID set*

consts
  ins :: *specID ⇒ chanID set*
  loc :: *specID ⇒ chanID set*
  out :: *specID ⇒ chanID set*

constdefs
  inStream :: *specID ⇒ chanID set ⇒ bool*
  inStream x y ≡ (*ins x = y*)

constdefs
  locStream :: *specID ⇒ chanID set ⇒ bool*
  locStream x y ≡ (*loc x = y*)

constdefs
  outStream :: *specID ⇒ chanID set ⇒ bool*
  outStream x y ≡ (*out x = y*)

constdefs
  correctInOutLoc :: *specID ⇒ bool*
  correctInOutLoc x ≡
  (*ins x*) ∩ (*out x*) = {}
   ∧ (*ins x*) ∩ (*loc x*) = {}
   ∧ (*loc x*) ∩ (*out x*) = {}

constdefs
  correctComposition :: *specID ⇒ bool*
  correctComposition x ≡
   subcomponents x = {} ⟶ *loc x* = {}

constdefs
  correctCompositionIn :: *specID ⇒ bool*
  correctCompositionIn x ≡
  (*ins x*) = (⋃ (*ins ' (subcomponents x*)) − (*loc x*))
   ∧ (*ins x*) ∩ (⋃ (*out ' (subcomponents x*))) = {}

constdefs
  correctCompositionOut :: *specID ⇒ bool*
  correctCompositionOut x ≡
  (*out x*) = (⋃ (*out ' (subcomponents x*))− (*loc x*))
   ∧ (*out x*) ∩ (⋃ (*ins ' (subcomponents x*))) = {}

constdefs
  correctCompositionLoc :: *specID ⇒ bool*
  correctCompositionLoc x ≡
  (*loc x*) = ⋃ (*ins ' (subcomponents x*)) ∩ ⋃ (*out ' (subcomponents x*))
end

The proof schema for the correctness properties above is standard and can be used automatically (if the proof fails, the specification of corresponding set is incorrect and must be changed) using the definition of the corresponding predicate and the Isabelle/HOL automatic proof strategy:

✓ For the property *correctInOutLoc*:

> **by** *(simp add: correctInOutLoc_def)*

✓ For the property *correctComposition*:

> **by** *(simp add: correctComposition_def)*

✓ for the property *correctCompositionIn*:

> **by** *(simp add: correctCompositionIn_def, auto)*

✓ for the property *correctCompositionOut*:

> **by** *(simp add: correctCompositionOut_def, auto)*

✓ for the property *correctCompositionLoc*:

> **by** *(simp add: correctCompositionLoc_def, auto)*

2.11.5. Example: Steam Boiler

In this section we start to present the case study "Steam Boiler". Fig. 2.3 presents an architecture of the system. In this section we discuss only the datatypes used to represent relations between sets of channels. The semantics of the components will be discussed later. In the specification group *SteamBoiler* will be used

✓ components *ControlSystem* (specification of requirements), *ControlSystemArch* (specification of system architecture presented on Fig. 2.3), *SteamBoiler*, *Converter*, and *Controller*;

✓ channels s, x, y and z.

The Isabelle/HOL specification of these types as well as component and channel identifiers is presented below by the theory *SteamBoiler_types.thy*. This theory does not contain any other type definitions, because the specification group *SteamBoiler* does not use any non-standard datatypes.

⚠ **Remark:** The theory *inout.thy* (see Section 2.11.4) must be copied into the same folder as the theory *stream.thy*, *SteamBoiler_types.thy*, *SteamBoiler_inout.thy* etc. The "SpecificationGroupName" must be replaced by the name of specification group – by *SteamBoiler*.

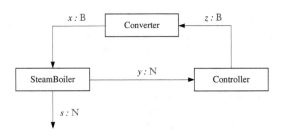

Figure 2.3.: Architecture of the Steam Boiler System

theory *SteamBoiler_types = Main + stream*:

datatype *specID =*
     sControlSystem
    | *sControlSystemArch*
    | *sSteamBoiler*
    | *sController*
    | *sConverter*

datatype *chanID =*
     ch_s
    | *ch_x*
    | *ch_y*
    | *ch_z*
end

The theory *SteamBoiler_inout.thy* is based only on the theory *inout.thy*. First of all we specify the subcomponent relations for all components of the system by the function *subcomponents*. Then we specify the list of input, output and local channels for all components by the functions *ins*, *out* and *loc* respectively. After that we prove that the predicates *correctInOutLoc* and *correctComposition* hold for all components, and also that the predicates *correctCompositionIn*, *correctCompositionOut* and *correctCompositionLoc* hold for the component *ControlSystemArch*, which is the only composite component of the specification group.

⚠ **Remark:** We do not need to use the Isabelle/HOL automatic proof strategy to prove that the predicates *correctCompositionIn* and *correctCompositionOut* hold for the component *ControlSystemArch*, because this component has no input channels. Thus, to prove these correctness properties it is enough to use simply the definition of the corresponding predicate.

theory *SteamBoiler_inout* = *Main* + *inout*:

primrec
  subcomponents sControlSystem   = {}
  subcomponents sControlSystemArch
     = {*sSteamBoiler, sController, sConverter*}
  subcomponents sSteamBoiler   = {}
  subcomponents sController = {}
  subcomponents sConverter = {}

primrec
  ins sControlSystem   = {}
  ins sControlSystemArch = {}
  ins sSteamBoiler   = {ch_x}
  ins sController = {ch_y}
  ins sConverter = {ch_z}

primrec
  loc sControlSystem   = {}
  loc sControlSystemArch = {ch_x, ch_y, ch_z}
  loc sSteamBoiler   = {}
  loc sController = {}
  loc sConverter = {}

primrec
  out sControlSystem   = {ch_s}
  out sControlSystemArch = {ch_s}
  out sSteamBoiler   = {ch_y, ch_s}
  out sController = {ch_z}
  out sConverter = {ch_x}

Proofs for components

ControlSystem:

lemma *spec_ControlSystem1*:
correctInOutLoc sControlSystem
  by (*simp add*: *correctInOutLoc_def*)

lemma *spec_ControlSystem2*:
correctComposition sControlSystem
  by (*simp add*: *correctComposition_def*)

  ControlSystemArch:

lemma *spec_ControlSystemArch1*:
correctInOutLoc sControlSystemArch
  by (*simp add*: *correctInOutLoc_def*)

lemma *spec_ControlSystemArch2*:
correctComposition sControlSystemArch
  by (*simp add*: *correctComposition_def*)

lemma *spec_ControlSystemArch3*:
correctCompositionIn sControlSystemArch
  by (*simp add*: *correctCompositionIn_def*)

lemma *spec_ControlSystemarch4*:
correctCompositionOut sControlSystemArch
  by (*simp add*: *correctCompositionOut_def*)

lemma *spec_ControlSystemArch5*:
correctCompositionLoc sControlSystemArch
  by (*simp add*: *correctCompositionLoc_def*, *auto*)

SteamBoiler:

lemma *spec_SteamBoiler1*:
correctInOutLoc sSteamBoiler
  by (*simp add*: *correctInOutLoc_def*)

lemma *spec_SteamBoiler2*:
correctComposition sSteamBoiler
  by (*simp add*: *correctComposition_def*)

Controller:

lemma *spec_Controller1*:
correctInOutLoc sController
  by (*simp add*: *correctInOutLoc_def*)

lemma *spec_Controller2*:
correctComposition sController
  by (*simp add*: *correctComposition_def*)

Converter:

lemma *spec_Converter1*:
correctInOutLoc sConverter
  by (*simp add*: *correctInOutLoc_def*)

lemma *spec_Converter2*:
correctComposition sConverter
  by (*simp add*: *correctComposition_def*)
end

2.12. Composition types

Composite specifications can be divided in the four types:

- ✓ sequential composition,
- ✓ parallel composition,
- ✓ "mix"-composition, and
- ✓ loops.

For all these types the specification semantics is defined according to Definition 1.3, but some of specification properties are different. Let us discuss now these types of composite specifications in more detail.

2.12.1. Sequential composition

The system *GraphSeqC* is a sequential composition of components C_1 and C_2. The component specifications C_1 and C_2 are represented by corresponding constraint on the communication histories of the channels. The FOCUS specification of the system in graphical style is given below.

The set of input channel identifiers of sequential composition of components C_1 and C_2 is equal to the set of input channels o_{C_1} of the component C_1, and set of output channels identifiers is equal to the set of output channels o_{C_2} of the component C_2.

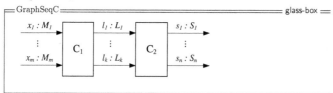

The semantics of the specification *GraphSeqC* is defined according to Definition 1.3 by

$$\llbracket GraphSeqC \rrbracket \overset{def}{=} \exists\, l_1 \in L_1^\infty, \dots l_k \in L_k^\infty : \llbracket C_1 \rrbracket \wedge \llbracket C_2 \rrbracket \qquad (2.38)$$

The specification *GraphSeqC* can be reformulated schematically (see [BS01]) into a specification *ConstrSeqC* in constraint style as shown below. The specifications *GraphSeqC* and *ConstrSeqC* have the same semantics:

$$\llbracket GraphSeqC \rrbracket = \llbracket ConstrSeqC \rrbracket$$

━━ ConstrSeqC ━━━━━━━━━━━━━━━━━━━━━━━━━━━━━━━━━━━━━ timed ━━

  in $x_1 : M_1;$ ...; $x_m : M_m$

  out $s_1 : S_1;$ ...; $s_n : S_n$

  loc $l_1 : L_1;$ ...; $l_k : L_k$

  $(l_1, \dots, l_k) := C_1(i_1, \dots, i_m)$
  $(s_1, \dots, s_n) := C_2(l_1, \dots, l_k)$

The representation in Isabelle/HOL will be exactly the definition of the semantics of a composite specification modulo Isabelle/HOL syntax:

$\exists\ l_1 :: L_1\ istream.\ \ldots\ \exists\ l_k :: L_k\ istream.$
$(c_1\ x_1\ \ldots\ x_m\ l_1\ \ldots l_k) \wedge (c_2\ l_1\ \ldots l_k\ s_1\ \ldots\ s_n)$
$\Longrightarrow\ constrSeqC\ x_1\ \ldots\ x_m\ s_1\ \ldots\ s_n$

2.12.2. Parallel composition

The system *GraphParC* is a parallel composition of components C_1 and C_2. The component specifications C_1 and C_2 are represented by corresponding constraint on the communication histories of the channels. The FOCUS specification[12] of the system in graphical style is given below.

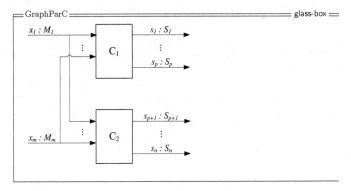

In general, the set of input channels identifiers of parallel composition of components C_1 and C_2 is a union (not always disjoint) of the sets of input channels i_{C_1} and i_{C_2} of the components C_1 and C_2 correspondingly. The system *GraphParC* shows the case, when all input streams of the component *GraphParC* are input streams of both components C_1 and C_2. It is also possible that the set of input channels identifiers of parallel composition of components C_1 and C_2 is a disjoint union of the sets of input channels i_{C_1} and i_{C_2}.

The set of output channels identifiers of parallel composition of components C_1 and C_2 is always a disjoint union of the sets of output channels o_{C_1} and o_{C_2} of the components C_1 and C_2 correspondingly. The semantics of the specification *GraphParC* is defined according to Definition 1.3 by

$$[\![GraphParC]\!] \stackrel{def}{=} [\![C_1]\!] \wedge [\![C_2]\!] \tag{2.39}$$

---

[12] The natural number p is here less than natural number n: $p < n$.

The specification *GraphParC* can be reformulated schematically into a specification *ConstrParC*, which has the same semantics as in the constraint style:

$$[\![GraphParC]\!] = [\![ConstrParC]\!]$$

═ ConstrParC ═══════════════════════════════════════ timed ═

  in $x_1 : M_1;$ ...; $x_m : M_m$

  out $s_1 : S_1;$ ...; $s_n : S_n$

$(s_1, \ldots, s_p) := C_1(x_1, \ldots, x_m)$
$(s_{p+1}, \ldots, s_n) := C_2(x_1, \ldots, x_m)$

The representation in Isabelle will be exactly the definition of semantics of composite specification modulo Isabelle/HOL syntax:

$$(c_1 \; x_1 \ldots x_m \; s_1 \ldots s_p) \wedge (c_2 \; x_1 \ldots x_m \; s_{p+1} \ldots s_n)$$
$$\Longrightarrow (constrParC \; x_1 \ldots x_m \; s_1 \ldots s_n)$$

The specification *GraphParC* (*ConstrParC*) can be time-synchronous only if both the specifications C_1 and C_2 are time-synchronous ones.

⚠ **Remark:** This kind of composition does not use any local channel.

2.12.3. "Mix"-composition

Let the component specifications C_1 and C_2 are represented by corresponding constraint on the communication histories of the channels. The system *GraphMixC* is a "mix"-composition of components C_1 and C_2. The FOCUS specification[13] of the system in graphical style is given below.

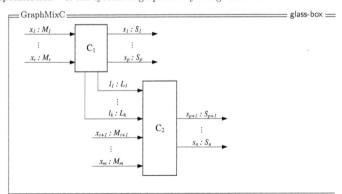

---

[13] The natural number p is here less than natural number n, and the natural number r is here less than natural number m: $p < n$ and $r < m$.

The semantics of the specification *GraphMixC* is defined according to Definition 1.3 by

$$\llbracket GraphMixC \rrbracket \stackrel{def}{=} \exists\, l_1 \in L_1^\infty, \ldots l_k \in L_k^\infty : \llbracket C_1 \rrbracket \wedge \llbracket C_2 \rrbracket \qquad (2.40)$$

The specification *GraphMixC* can be reformulated schematically into a specification *ConstrMixC*, which has the same semantics as in constraint style:

$$\llbracket GraphMixC \rrbracket = \llbracket ConstrMixC \rrbracket$$

ConstrMixC ———————————————————————————— timed ——

in $x_1 : M_1;$ $\ldots;$ $x_m : M_m$

out $s_1 : S_1;$ $\ldots;$ $s_n : S_n$

loc $l_1 : L_1;$ $\ldots;$ $l_k : L_k$

$(s_1, \ldots, s_p) :- C_1(x_1, \ldots, x_r)$

$(s_{p+1}, \ldots, s_n) := C_2(l_1, \ldots, l_k, x_{r+1}, \ldots, x_m)$

The representation in Isabelle will be exactly the definition of semantics of composite specification modulo Isabelle/HOL syntax:

$\exists\, l_1 :: L_1 \; istream. \; \ldots \; \exists\, l_k :: L_k \; istream.$
$(c_1 \; x_1 \; \ldots \; x_r \; l_1 \; l_k \; s_1 \; \ldots s_p) \wedge (c_2 \; l_1 \; \ldots l_k \; x_{r+1} \; \ldots x_m \; s_{p+1} \; \ldots \; s_n)$
$\Longrightarrow constrMixC \; x_1 \; \ldots \; x_m \; s_1 \; \ldots \; s_n$

2.12.4. Loop

Let the component specification C is represented by corresponding constraint on the communication histories of the channels. When a component *GraphLoopC* is build from a component C connecting its output channel with input one, this channel makes a communication loop. The Focus specification *GraphLoopC* of a system in graphical style is given below.

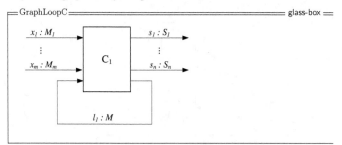

The semantics of the specification *GraphLoopC* is defined according to Definition 1.3 by

$$[\![GraphLoopC]\!] \overset{def}{=} \exists\, l_1 \in M^{\infty} : [\![C_1]\!] \tag{2.41}$$

The specification *GraphLoopC* can be reformulated schematically into a specification *ConstrLoopC*, which has the same semantics as in the specification in constraint style:

$$[\![GraphLoopC]\!] = [\![ConstrLoopC]\!]$$

ConstrLoopC ━━━━━━━━━━━━━━━━━━━━━━━━━━━━━ timed ━━

in $x_1 : M_1;$ ...; $x_m : M_m$

out $s_1 : S_1;$ ...; $s_n : S_n$

loc $l_1 : M$

$(s_1, \ldots, s_n, l_1) := C(x_1, \ldots, x_r, l_1)$

If the specification of the C component is only *weakly* causal, this leads to Zeno paradoxes. To prevent Zeno paradoxes some delays for the "loop"-channels are needed (in each time interval only *finite* sequence of messages can be presented on a channel). Thus, for the "loop"-composition the original component must be *strong* causal (with delays) and an additional initial value of the connected stream must be specified. The component specification *ConstrLoopCInit* is an extension of the *ConstrLoopC* by using an initial value *InitValue* of the connected stream:

ConstrLoopCInit ━━━━━━━━━━━━━━━━━━━━━━━━━ timed ━━

in $x_1 : M_1;$ ...; $x_m : M_m$

out $s_1 : S_1;$ ...; $s_n : S_n$

loc $l_1 : M$

$(s_1, \ldots, s_n, l_1) := C(x_1, \ldots, x_r, l_2)$
where l_2 so that
   $\mathsf{ti}(l_2, 1) = \langle InitValue \rangle$
   $\mathsf{ti}(l_2, i+1)) = \mathsf{ti}(l_1, i)$

We can also see such a composed specification as a composition of two components C and *InitV*, as shown in the specification *GraphLoopCInitV*:

$$[\![GraphLoopCInitV]\!] = [\![ConstrLoopCInit]\!]$$

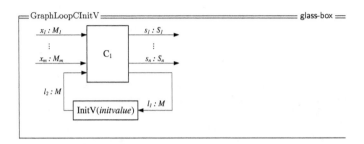

We specify the component *InitV* as a parameterized one – the parameter is an initial value of a stream, more precisely, the list of messages that must be produced at the first time interval.

╔═ InitV $(InitValue \in M^*)$ ═══════════════════════════════ timed ═
in $l_1 : M$

out $l_2 : M$
───
$\text{ti}(l_2, 1) = \langle InitValue \rangle$
$\text{ti}(l_2, i + 1) = \text{ti}(l_1, i)$

The corresponding specifications in constraint style are given below:

╔═ ConstrLoopCInitV ═══════════════════════════════════════ timed ═
in $x_1 : M_1; \quad \ldots; \quad x_m : M_m$

out $s_1 : S_1; \quad \ldots; \quad s_n : S_n$

loc $l_1 : M; \quad l_2 : M$
───
$(s_1, \ldots, s_n, l_1) := C(x_1, \ldots, x_m, l_2)$
$(l_1) := \text{InitV}(InitValue)(l_2)$

The representation in Isabelle will be exactly the definition of semantics of composite specification modulo Isabelle/HOL syntax:

$\exists l_1 :: L_1 \text{ istream. } (c \ x_1 \ \ldots \ x_m \ l_2 \ s_1 \ \ldots s_n \ \dot{l_2}) \land (initV \ l_2 \ l_1)$
$\implies constrLoopCInit \ x_1 \ \ldots \ x_m \ s_1 \ \ldots \ s_n$

⚠ **Remark:** If the delay between input and output streams in the main component (in this case, in the component *GraphLoopCInit*) is greater than one, i.e. some natural number d, $1 < d$, then the initial values must be defined not only for the first time interval, but for the first d time intervals.

───

2.13. Sheaves and Replications

As mentioned in Section 1.3.5, a specified system can contain a number of copies of channel of the same type or several instances of the same component. If this number of copies is finite, fixed and small enough, we can use the simple composition kinds, which are introduced in Section 2.11. But if the number of copies must be specified as some variable of type $\mathbb{N}$ or if the number of copies is finite and fixed, but too large to have a readable system specification, the notions of *sheaf of channels* and *replication of specifications* must be used (see Section 1.3.5).

2.13.1. Sheaves of Channels

In addition to the notation presented in Section 1.3.5

$$x[\{1, \ldots, k\}] \ : \ M,$$

the following notation can also be used to specify a sheaf of channels in FOCUS:

$$x_1, \ldots, x_k \ : \ M.$$

From here we prefer the notation $x_1, \ldots, x_k \ : \ M$.

We say that a sheaf of channels $x_1, \ldots, x_n$ is *correct*, if all the channels $x_1, \ldots, x_n$ are of the same type and the number n is greater than zero.

To represent in Isabelle/HOL a sheaf of timed infinite streams $x_1, \ldots, x_n$ of some type *Streamtype* we propose to use the following kind of functional types:

types *nStreamtype* $=$ *nat* $\Rightarrow$ *streamtype istream*

A sheaf will be specified in Isabelle/HOL as a single variable of corresponding type, e.g. the sheaf $x_1, \ldots, x_n$ will be represented as a variable nX of type *nStreamtype*. To translate the FOCUS formula over channels (streams) from a sheaf, e.g. to say that the predicate p is true for any stream of the sheaf $send_1, \ldots, send_n$ (in FOCUS this formula is represented by $\forall\, i \in [1..n] : \ p(s_i)$) the following notation can be used[14]:

$\forall\, i < n.\ p\ (nSend\ i)$

To argue about sheaves of channels in Isabelle/HOL we need to make sure that the sheaf is nonempty[15]. For this purpose the Isabelle/HOL predicate *CorrectSheaf n* is used. This predicate is true, if the number n of channels is greater than zero. From now on we will refer to this number n as *sheaf upper bound*.

---

[14] The relation $<$ must be used, because the elements in Isabelle/HOL are counted from 0, in contrast to FOCUS, where the count goes from 1.

[15] In FOCUS this is automatically true: the notation $x_1, \ldots, x_n$ implies that $0 < n$.

The Isabelle/HOL semantics of a Focus specification that uses sheaf(s) of channels will be specified in the same manner as for a specification without sheaves of channels (see Section 2.11). The only extension is that the predicate(s) *CorrectSheaf* over the corresponding sheaf(s) of streams must be added as extra conjunct(s) to the definition of the specification semantics in Isabelle/HOL (see Equations 2.23 and 2.29).

The Isabelle/HOL specification of relation between sets of channels and sheaves of channels is more complicated than for "simple" channels. The correctness proofs are also more complicated, but still be schematic. We will discuss this difference in Section 2.13.3.

⚠ **Remark:** Lists of specification parameters of the same type will be represented in a similar way as sheaves of channels. In this case we can speak of a *sheaf of parameters*.

In more general case[16], when the set *Id* of indexes in the sheaf is not a subset of the set of natural numbers, e.g. $Id = \{a, b, c, d\}$, we understand by the *sheaf upper bound* the whole set *Id*.

First of all we need to specify the type of elements of the set *Id*:

  datatype $Id = a \mid b \mid c \mid d$

To represent a sheaf $x[Id]$ of timed infinite streams of some type *Streamtype* in Isabelle/HOL we will use the following kind of functional types:

  types $nStreamtype = Id\ set \Rightarrow streamtype\ istream$

A sheaf of this kind will be specified in Isabelle/HOL also as a single variable of corresponding type, e.g. the sheaf $x[Id]$ will be represented as a variable nX of type *nStreamtype*. To translate the Focus formula over channels (streams) from a sheaf, e.g. to say that the predicate p is true for any stream of the sheaf $x[Id]$ ($\forall i \in Id : p(x_i)$), a similar notation will be used:

  $\forall (i :: IdSet).\ p\ (x\ i)$

The predicate *CorrectSheaf* is not needed in this case, because the set (type) *Id* is nonempty one.

Hence, the Isabelle/HOL predicate that represent the semantics of a Focus specification with sheaf(s) of channels must have an extra parameter for each sheaf upper bound. For example, the following four sheaves of channels in some specification

- ✓ $a_1, \ldots, a_n$,
- ✓ $b_1, \ldots, b_m$,
- ✓ $c_1, \ldots, c_n$, and
- ✓ $d[i], i \in Id$

correspond to three upper bounds m, n, and *Id*. Thus, the Isabelle/HOL predicate that represent the semantics of this specification must have three extra parameters: two of type *nat* and one of type *Id*.

---

[16]This case implies often less readability of a specification.

Example 2.6:

The Focus specification of FlexRay communication protocol requirements (for details see Section 4.2) is an example of a system, that requires to use the notion of sheaf of channels.

FlexRay (constant $c_1, ..., c_n \in$ Config) ━━━━━━━━━━━━━━━━━ timed ━

in $return_1, ..., return_n : Frame$

out $store_1, ..., store_n : Frame;\ get_1, ..., get_n : \mathbb{N}$

---

asm $\forall\, i \in [1..n] : \mathsf{msg}_1(return_i)$

        $DisjointSchedules(c_1, ..., c_n)$

        $IdenticCycleLength(c_1, ..., c_n)$

gar $FrameTransmission(return_1, ..., return_n, store_1, ..., store_n, get_1, ..., get_n,$

          $c_1 ..., c_n)$

        $\forall\, i \in [1..n] : \mathsf{msg}_1(get_i) \wedge \mathsf{msg}_1(store_i)$

In the specification *FlexRay* we have four sheaves of channels: $store_1, ..., store_n$ (of type *Frame*), $get_1, ..., get_n$ (of type $\mathbb{N}$), $return_1, ..., return_n$ (of type *Frame*) and also a list of parameters $c_1, ..., c_n$ (of type *Config*). To represent them we define in Isabelle/HOL the special types *nFrame*, *nNat*, and *nConfig*. For example, the type *nNat* is used to represent sheaf of channels of type $\mathbb{N}$:

    types *nNat* $=$ *nat* $\Rightarrow$ *nat istream*

The representation of the Focus specification *FlexRay* of the communication protocol requirements in Isabelle/HOL is give below. The whole Focus specifications and its translations to Isabelle/HOL together with corresponding proofs is discussed in details in Section 4.2.

constdefs
  FlexRay ::
    nat $\Rightarrow$ *'a nFrame* $\Rightarrow$ *nConfig* $\Rightarrow$ *'a nFrame* $\Rightarrow$ *nNat* $\Rightarrow$ *bool*
  FlexRay n nReturn nC nStore nGet
  $\equiv$
  (*CorrectSheaf n* $\wedge$
  ($\forall\ i < n.\ maxmsg\ 1\ (nReturn\ i)$) $\wedge$
  (*DisjointSchedules n nC*) $\wedge$ (*IdenticCycleLength n nC*)
  $\longrightarrow$
  ((*FrameTransmission n nReturn nStore nGet nC*) $\wedge$
  ($\forall\ i < n.\ maxmsg\ 1\ (nGet\ i) \wedge maxmsg\ 1\ (nStore\ i)$))))

$\square$

2.13.2. Specification Replications

Let us recall the specification *RepC* from Section 1.3.5. To make a specification more readable and the corresponding proofs simpler, we take the set *Cid* as a subset of natural numbers

$$Cid = \{1, \ldots, N\}$$

where N is the number of replications. For such a case is more suited the "uncompressed" graphical representation:

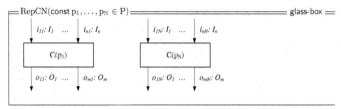

Hence the semantics of the specifications *RepC* and *RepCN* is equal, they have the same Focus representation as plain text (constraint style):

$$\boxed{\begin{array}{l} \text{RepCN(const } p_1, \ldots, p_N \in P) \hspace{3cm} \text{timed} \\[4pt] \text{in} \quad i_{11} : I_1, \ldots i_{1N} : I_1; \ldots; i_{n1} : I_n, \ldots i_{nN} : I_n \\[4pt] \text{out} \quad o_{11} : O_1, \ldots o_{1N} : O_1; \ldots; o_{m1} : O_m, \ldots o_{mN} : O_m \\[8pt] \forall j \in [1..N] : \\ \quad (o_{1j}, \ldots, o_{mj}) := C(p_j)(i_{1j}, \ldots, i_{nj}) \end{array}}$$

⚠ **Remark:** The advantage of the "uncompressed" representation is a better readability for the cases when we have in a specification not only sheaves of channels, but also some single channel which is an input for all replications.

The Isabelle/HOL representation of a specification replication is based on the representation of sheaves of channels and parameters:

```
constdefs
  RepCN ::
    nat ⇒ nI1 ⇒ ... ⇒ nIn ⇒ nP ⇒ nO1 ⇒ ... ⇒ nOm ⇒ bool
  RenCN n ni1 ... nin nP nO1 ... nOm ≡
  (CorrectSheaf n ∧
   (∀ i < n. C (ni1 i) ... (nin i) (nP i) (nO1 i) ... (nOm i) ))
```

The Isabelle/HOL notation

$$\forall i < n. \; SomePredicate$$

---

means in the internal Isabelle/HOL representation

$$\forall\ i.\ i < n \longrightarrow SomePredicate$$

In the case we do not take for some reasons $Cid = \{1, \dots, N\}$ for some $N \in \mathbb{N}$, but define some datatype Id (see the previous section for details), the representation will be similar one:

constdefs
  RepCN ::
    Id set $\Rightarrow$ nI1 $\Rightarrow$ ... $\Rightarrow$ nIn $\Rightarrow$ nP $\Rightarrow$ nO1 $\Rightarrow$ ... $\Rightarrow$ nOm $\Rightarrow$ bool
  RenCN IdSet ni1 ... nin nP nO1 ... nOm $\equiv$
    ($\forall$ (i :: IdSet). C (ni1 i) ... (nin i) (nP i) (nO1 i) ... (nOm i))

 Remark: The concepts of sheaves and replications can be used for the JANUS specifications in the same manner as for the FOCUS specifications. The corresponding representation in Isabelle/HOL will be the same.

Example 2.7:
Specifications *FlexRayArchitecture* and *FlexRayArchitectureR* represent the same system (the guarantee part architecture of the FlexRay, for details see Section 4.2). They are semantically equal, but they use different kinds of replication representation: in the specification *FlexRayArchitecture* the "compressed" kind of representation was used, where the specification *FlexRayArchitectureR* is presented in the "uncompressed" kind. □

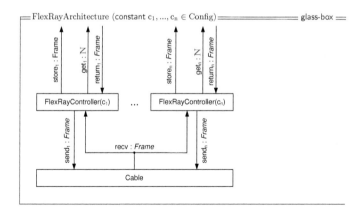

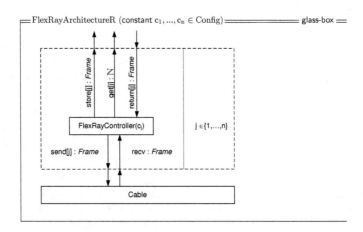

2.13.3. Isabelle/HOL Specification of Relations between Sets of Channels

To represent the Isabelle/HOL semantics of the sets of input, output and local channel and sheaf identifiers of a FOCUS specification S, we need to define first of all the following datatypes:

✓ *chanID* – the type of all needed channel indetifiers $Id_1, \ldots, Id_N$:

$$\textbf{\textit{datatype}} \ \ chanID = ch\_Id_1 \mid \cdots \mid ch\_Id_N$$

✓ *specID* – the type of all needed specification names $S_1, \ldots, S_M$ (specification identifiers):

$$\textbf{\textit{datatype}} \ \ specID = sS_1 \mid \cdots \mid sS_M$$

After that we add the definition of

✓ *csID* – the joint type of channel and sheaves identifiers, as well as

✓ *spID* – the joint type of specification and specification replications identifiers:

datatype *csID* = *ch chanID* | *sheaf csID nat*

datatype *spID* = *spec specID* | *repl spID nat*

These types must be defined for the specification group as whole in the theory *SpecificationGroupName_types.thy* (see Sections 2.13.5). In this theory we also need to declare variables which represents the upper bounds of the sheaves and replications and specify the corresponding list *sheafNumbers* of all upper bounds, used in the specification group, e.g.

consts sN :: nat

constdefs
  sheafNumbers :: *nat list*
  sheafNumbers $\equiv [sN]$

If a local stream x in a composite specification S is split into a sheaf of channels, we need to add this information to the Isabelle/HOL theory *SpecificationGroup-Name_types.thy* to argue about the composition correctness. We represent this information by the schematically defined predicate *ch_split* over the types *csID* and *nat*. The predicate *ch_split* holds for a channel identifier c and a natural number n, if the channel (stream) c is split in a sheaf with upper bound n, e.g. (for a detailed example see Sections 2.13.5 and 4.2):

constdefs
  ch_split :: $csID \Rightarrow nat \Rightarrow bool$
  ch_split $x\ l \equiv$
    $(x = ch\ ch\_recv \wedge l = sN)$

To argue about the composition correctness in the case of split channel, we also need to define the following auxiliary predicates:

✓ *eqSplit*: to specify the relation between two sets of sheaf or channels identifiers, which represents equality modulo split channels;

✓ *chan2Sheaf*: to convert a channel identifier to a sheaf identifier with a given upper bound;

✓ *cs2Sheaf*: to convert a channel or a sheaf identifier to a sheaf identifier with a given upper bound (this predicate can also be used to build sheaves of sheaves of channels);

✓ *makeSheafs*: to build set of sheaf identifiers from a channel or a sheaf identifier and the given upper bound (based on the definition of the predicate *ch_split*);

✓ *split2sheaf*: to build set of sheaf identifiers from a channel or a sheaf identifier and the list of sheaf upper bounds *sheafNumbers* (based on the definition of the predicate *makeSheafs*);

✓ *extSplit*: to extend a set of channel or sheaf identifiers by the sheaf identifiers, which have been got by splitting;

✓ *split2loc*: to convert a channel or a sheaf identifier into a set, where an identifier of a sheaf, which has been got by splitting, is replaced by the corresponding channel identifier.

In the similar manner as for specifications without sheaves and replications we specify the sets of input, output and local identifiers of the specification S as functions *ins*, *out* and *loc*, which return for a given specification name the list of corresponding channel *or sheaf* identifiers – input, output and local respectively.

Then we specify extensions of these functions to work also with specification replication identifiers: the functions *insS*, *outS* and *locS* returns for a given specification *or specification replication* name the list of corresponding channel *or sheaf* identifiers – input, output and local respectively.

We also introduce predicates *inStream*, *outStream* and *locStream* over specification *or specification replication* names and channel *or sheaves* identifiers. The predicate *inStream S x* (*outStream S x*, *locStream S x*) holds, if the list of identifiers x corresponds to the specification (replication) name S.

We specify the function *subcomponents* which returns for a given component the list of its subcomponents in the same way as for specification group without replications. The analog defined function *subcomponentS* returns the list of subcomponents for a given component or component replication.

The definitions of the functions *insS*, *outS*, *locS* and *subcomponentS* are independent from the concrete system specification. They can be seen as a part of the schema and simply copied without any changes to the theory *Specification-GroupName_inout.thy*.

The predicates to express the correctness properties (*correctInOutLoc*, *correct-Composition*, *correctCompositionIn*, *correctCompositionOut*, and *correctComposition-Loc*) are defined similar to the case of specification groups without sheaves and replications (see Section 2.11.4), but now they are specified over the set *spID* and taking into account the possibility that some channels are split into sheaves.

⚠ **Remark:** The theory *inout_sheaf.thy* is independent of the way the the sheaves of channels are defined – over the set ℕ with some variables as the sheaves upper bound, or over some extra defined finite nonempty set.

theory *inout_sheaf* = *Main* + *FR_types*:

consts
  subcomponents :: *specID* ⇒ *spID set*
consts
  subcomponentS :: *spID* ⇒ *spID set*

consts
  ins :: *specID* ⇒ *csID set*
  loc :: *specID* ⇒ *csID set*
  out :: *specID* ⇒ *csID set*

consts
  insS :: *spID* ⇒ *csID set*
  locS :: *spID* ⇒ *csID set*
  outS :: *spID* ⇒ *csID set*

constdefs
  chan2Sheaf :: *chanID* ⇒ *nat* ⇒ *csID*
  chan2Sheaf x i ≡ *sheaf* (*ch x*) *i*

constdefs
  cs2Sheaf :: *nat* ⇒ *csID* ⇒ *csID*
  cs2Sheaf i x ≡ *sheaf x i*

constdefs
  inStream :: *spID* ⇒ *csID set* ⇒ *bool*
  inStream x y ≡ (*insS x* = *y*)

constdefs
  locStream :: *spID* ⇒ *csID set* ⇒ *bool*
  locStream x y ≡ (*locS x* = *y*)

constdefs
  outStream :: *spID* ⇒ *csID set* ⇒ *bool*
  outStream x y ≡ (*outS x* = *y*)

constdefs
  eqSplit :: *csID set* ⇒ *csID set* ⇒ *bool*
  eqSplit x y
    ≡
    ($\forall$ *i* ∈ *x*. *i* ∈ *y* ∨ ($\exists$ (*n::nat*).(*ch_split i n*) ∧ (*sheaf i n*) ∈ *y*)) ∧
    ($\forall$ *j* ∈ *y*. *j* ∈ *x* ∨ ($\exists$ *c* ∈ *x*. $\exists$ (*n::nat*).
                  (*sheaf c n*) = *j* ∧ (*ch_split c n*)))

constdefs
  correctInOutLoc :: *spID* ⇒ *bool*
  correctInOutLoc x ≡
    (*insS x*) ∩ (*outS x*) = {}
    ∧ (*insS x*) ∩ (*locS x*) = {}
    ∧ (*locS x*) ∩ (*outS x*) = {}

constdefs
  correctComposition :: *spID* ⇒ *bool*
  correctComposition x ≡
    subcomponentS x = {} ⟶ *locS x* = {}

constdefs
  makeSheafs :: *csID* ⇒ *nat* ⇒ *csID set*
  makeSheafs x n ≡
    if (*ch_split x n*)
    then {*sheaf x n*}
    else {}

constdefs
  split2sheaf :: *csID* ⇒ *csID set*
  split2sheaf x ≡
    (*if* (∃ *n.* (*n mem sheafNumbers*) ∧ *ch_split x n*)
    then ⋃ ((*makeSheafs x*) ' (*set sheafNumbers*))
    else {})

constdefs
  extSplit :: *csID set* ⇒ *csID set*
  extSplit x ≡ *x* ∪ ⋃ (*split2sheaf* ' *x*)

constdefs
  correctCompositionIn :: *spID* ⇒ *bool*
  correctCompositionIn x ≡
  eqSplit (*insS x*) (⋃ (*insS* ' (*subcomponentS x*)) − *extSplit* (*locS x*))
  ∧ (*insS x*) ∩ ⋃ (*outS* ' (*subcomponentS x*)) = {}

constdefs
  correctCompositionOut :: *spID* ⇒ *bool*
  correctCompositionOut x ≡
  eqSplit (⋃ (*outS* ' (*subcomponentS x*)) − *extSplit* (*locS x*)) (*outS x*)
  ∧ (*outS x*) ∩ ⋃ (*insS* ' (*subcomponentS x*)) = {}

constdefs
  split2loc :: *csID* ⇒ *csID set*
  split2loc x ≡
    (*case x of*
      ch i ⇒ {*x*}
    | *sheaf j n* ⇒ (*if ch_split j n*
            then {*j*}
            else {*x*}))

constdefs
  correctCompositionLoc :: *spID* ⇒ *bool*
  correctCompositionLoc x ≡
  eqSplit (*locS x*)
        (⋃ ((*split2loc* ' (⋃ (*insS* ' (*subcomponentS x*))))
           ∩ (*split2loc* ' (⋃ (*outS* ' (*subcomponentS x*))))))
end

The Isabelle/HOL theory *inout_sheaf.thy*, like the theory *inout.thy* for specification groups without sheaves and replications, uses the (user-defined) theory *SpecificationGroupName_types.thy*, which contains all user-defined datatypes (including *specID* and *chanID*, as well as schematically defined *scID* and *spID*) used in the specification group. The theory *inout_sheaf.thy* must be copied into the same folder as the theory *stream.thy*, *SpecificationGroupName_types.thy*, *SpecificationGroupName_inout.thy* etc. "SpecificationGroupName" must be replaced by the actual name of specification group.

The proof schema for the case of specification replications and sheaves of channels is also standard: [17]

✓ for the property *correctInOutLoc*:

> **by** *(simp add: correctInOutLoc_def)*

✓ for the property *correctComposition*:

> **by** *(simp add: correctComposition_def)*

✓ for the property *correctCompositionIn*:

> **by** *(simp add: correctCompositionIn_def extSplit_def*
> *ch_split_def sheafNumbers_def eqSplit_def*
> *cs2Sheaf_def split2sheaf_def makeSheafs_def,*
> *auto)*

In some cases applying of the Isabelle/HOL automatic proof strategy can be omitted. For the special case, where a composite specification has no local channels, it is enough to use the definitions of the predicates *correctCompositionIn*, *extSplit* and *eqSplit*.

✓ for the property *correctCompositionOut*:

> **by** *(simp add: correctCompositionOut_def*
> *extSplit_def ch_split_def eqSplit_def*
> *cs2Sheaf_def split2sheaf_def makeSheafs_def,*
> *auto)*

For the special case, where a composite specification has no local channels, it is enough to use the definitions of the predicates *correctCompositionOut*, *extSplit* and *eqSplit*.

✓ for the property *correctCompositionLoc*:

> **by** *(simp add: correctCompositionLoc_def*
> *eqSplit_def ch_split_def split2loc_def cs2Sheaf_def, auto)*

For the special case, where a composite specification has no local channels, it is enough to use the definitions of the predicates *correctCompositionLoc*, *eqSplit* and *split2loc*.

---

[17]The proofs of the properties *correctInOutLoc* and *correctComposition* are the same as for the specifications without replications and sheaves of channels.

2.13.4. Disjoint Channels in a Sheaf

In some cases we need to have such a sheaf of channels of some type M (with the identifier set $IdSet$), where at most one channel is "active" per time interval, i.e. at every time interval at most one channel contains some messages. For this reasons we define FOCUS operator

$$\mathsf{disj}_\mathsf{s}^\mathsf{inf} \in M^\infty[IdSet] \rightarrow \mathbb{Bool}$$

$$\mathsf{disj}_\mathsf{s}^\mathsf{inf}(s[IdSet]) \stackrel{\mathsf{def}}{=}$$
$$\forall\, t \in \mathbb{N} : \forall\, i, j \in IdSet, i \neq j : \mathsf{ti}(s[i], t) \neq \langle\rangle \rightarrow \mathsf{ti}(s[j], t) = \langle\rangle$$

We translate the FOCUS operator $\mathsf{disj}_\mathsf{s}^\mathsf{inf}$ into the Isabelle/HOL predicate *inf_disjS* (the type variables $'a$ corresponds here to the polymorphic types M and $IdSet$ respectively):

```
constdefs
  inf_disjS :: 'b set ⇒ ('b ⇒ 'a istream) ⇒ bool
  inf_disjS IdSet nS
  ≡
  ∀ (t::nat) i j. (i:IdSet) ∧ (j:IdSet) ∧    ((nS i) t) ≠ []
    ⟶ ((nS j) t) = []
```

For the case $IdSet = \{1..n\}$, $n \in \mathbb{N}$ we can represent the FOCUS operator $\mathsf{disj}_\mathsf{s}^\mathsf{inf}$ in Isabelle/HOL as follows:

$$\mathsf{disj}^\mathsf{inf} \in M^\infty \times \cdots \times M^\infty \rightarrow \mathbb{Bool}$$

$$\mathsf{disj}^\mathsf{inf}(s_1, \ldots, s_n) \stackrel{\mathsf{def}}{=}$$
$$\forall\, t \in \mathbb{N} : \forall\, i, j \in [1..n], i \neq j : \mathsf{ti}(s[i], t) \neq \langle\rangle$$
$$\rightarrow \mathsf{ti}(s[j], t) = \langle\rangle$$

We translate the FOCUS operator $\mathsf{disj}^\mathsf{inf}$ into the Isabelle/HOL predicate *inf_disj* (the type variable $'a$ corresponds here to the polymorphic type M):

```
constdefs
  inf_disj :: nat ⇒ (nat ⇒ 'a istream) ⇒ bool
  inf_disj n nS
  ≡
  ∀ (t::nat) (i::nat) (j::nat).
    i < n ∧ j < n ∧ i ≠ j ∧ ((nS i) t) ≠ []   ⟶
    ((nS j) t) = []
```

2.13.5. Example: System with Sheaves of Channels

We discuss an example of a system with sheaf of channels. For this system we have two specifications – *KReq*, a specification of system requirements, and *K*, a specification of a system architecture. The specification *K* is a composite one. The subcomponents of *K* are components *A*, B_1, ..., B_n, and *F*. Specifications of these components are elementary ones.

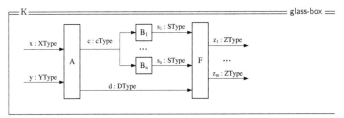

Thus, the specification group *Group* has specifications of five kinds: *KReq*, *K*, *A*, *B*, and *F*, and channels of six kinds: *x*, *y*, *c*, *d*, *s*, and *z*.

In the specification *K* the stream *c* is split a sheaf of channels. Therefore, we need to add this information to the Isabelle/HOL specification *Group_types.thy* defining the *ch_split* predicate.

The constant *sN* is declared for the number of channels in sheaves (specification in replications).

⚠ **Remark:** If the corresponding types *XType*, *YType*, *CType*, *SType*, and *ZType* are not basic ones and do not inherit from some other predefined theories, they also need to be defined in this theory. In the case we inherit some type of a theory *T* we need to add its name to the theory header.

The Isabelle/HOL theory *Group_inout.thy* is based only on the Isabelle/HOL theory *inout_sheaf.thy* (see Section 2.13.3), which is in the case of the specification group *Group* based on the theory *Group_types.thy*. First of all we specify in this theory the subcomponent relations for all components of the system by the function *subcomponents*. Then we specify the list of input, output and local channels for all components by the functions *ins*, *out* and *loc* respectively.

theory *Group_types = Main + stream*:

consts *sN :: nat*

constdefs
sheafNumbers :: nat list
sheafNumbers ≡ [sN]

datatype *chanID =*
   ch_x
  | *ch_y*
  | *ch_c*
  | *ch_s*
  | *ch_d*
  | *ch_z*

datatype *specID =*
   sKReq
  | *sK*
  | *sA*
  | *sB*
  | *sF*

datatype *csID = ch chanID | sheaf csID nat*
datatype *spID = spec specID | repl spID nat*

constdefs
ch_split :: csID ⇒ nat ⇒ bool
ch_split x n ≡ (x = ch ch_c ∧ n = sN)
end

theory *Group_inout = Main + inout_sheaf*:

primrec
subcomponents sKReq = {}
subcomponents sK = {spec sA, repl (spec sB) sN, spec sF}
subcomponents sA = {}
subcomponents sB = {}
subcomponents sF = {}

primrec
subcomponentS (spec x) = subcomponents x
subcomponentS (repl x y) = subcomponentS x

primrec
ins sKReq = {ch ch_x, ch ch_y}
ins sK = {ch ch_x, ch ch_y}
ins sA = {ch ch_x, ch ch_y}
ins sB = {ch ch_c}
ins sF = {sheaf (ch ch_s) sN, ch ch_d}

primrec
 insS (*spec x*)   = *ins x*
 insS (*repl x i*) = (*cs2Sheaf i · (insS x)*)

primrec
 loc sKReq = {}
 loc sK = {*ch ch_c, sheaf* (*ch ch_s*) *sN, ch ch_d*}
 loc sA = {}
 loc sB = {}
 loc sF = {}

primrec
 locS (*spec x*)   = *loc x*
 locS (*repl x i*) = (*cs2Sheaf i · (locS x)*)

primrec
 outS (*spec x*)   = *out x*
 outS (*repl x i*) = (*cs2Sheaf i · (outS x)*)

primrec
 out sKReq = {*sheaf* (*ch ch_z*) *sN*}
 out sK   = {*sheaf* (*ch ch_z*) *sN*}
 out sA = {*ch ch_c, ch ch_d*}
 out sB = {*ch ch_s*}
 out sF = {*sheaf* (*ch ch_z*) *sN*}

Proofs for components

lemma *spec_KReq1*:
correctInOutLoc (*spec sKReq*)
 by (*simp add: correctInOutLoc_def*)

lemma *spec_KReq2*:
correctComposition (*spec sKReq*)
 by (*simp add: correctComposition_def*)

lemma *spec_K1*:
correctInOutLoc (*spec sK*)
 by (*simp add: correctInOutLoc_def*)

lemma *spec_K2*:
correctComposition (*spec sK*)
 by (*simp add: correctComposition_def*)

lemma *spec_K3*:
correctCompositionIn (*spec sK*)
 by (*simp add: correctCompositionIn_def*
    extSplit_def eqSplit_def split2sheaf_def
    cs2Sheaf_def sheafNumbers_def ch_split_def
    makeSheafs_def)

lemma *spec_K4* :
correctCompositionOut (spec sK)
  by (*simp add: correctCompositionOut_def*
    extSplit_def ch_split_def eqSplit_def
    split2sheaf_def makeSheafs_def cs2Sheaf_def,
    auto)

lemma *spec_K5* :
correctCompositionLoc (spec sK)
    by (*simp add: correctCompositionLoc_def ch_split_def*
      eqSplit_def split2loc_def cs2Sheaf_def,
      auto)

lemma *spec_A1* :
correctInOutLoc (spec sA)
    by (*simp add: correctInOutLoc_def*)

lemma *spec_A2* :
correctComposition (spec sA)
  by (*simp add: correctComposition_def*)

lemma *spec_B1* :
correctInOutLoc (spec sB)
  by (*simp add: correctInOutLoc_def*)

lemma *spec_B2* :
correctComposition (spec sB)
  by (*simp add: correctComposition_def*)

lemma *spec_F1* :
correctInOutLoc (spec sF)
  by (*simp add: correctInOutLoc_def*)

lemma *spec_F2* :
correctComposition (spec sF)
  by (*simp add: correctComposition_def*)

end

2.14. Translation schema: From FOCUS to Isabelle/HOL

The transition schema for the FOCUS types was given in Section 2.2. The translation of a FOCUS specification into Isabelle/HOL was presented in general in Section 2.11 by Equations 2.23 (elementary specification) and 2.33 (composite specification). Representation of sheaf channels and specification replications was discussed in Section 2.13.

The body of an elementary specification can be represented in FOCUS in the following ways:

✓ tiTable;

✓ state transition diagram;

✓ a number of properties defined as logic formulas.

The translation of a FOCUS tiTable into Isabelle/HOL was presented in Section 2.6, and the representation of a state transition diagram was given in Section 2.8. In this section we summarise the results of Sections 2.2 und 2.5 (representation of the FOCUS operators in Isabelle/HOL) in Tables 2.4–2.7 and discuss the transition schema for the FOCUS types, expressions, function and predicates.[18] The transition schema for the FOCUS formulas is given in Table 2.8.

FOCUS operator, f	Isabelle/HOL representation, $[\![f]\!]_{Isab}$
$\langle\rangle$	$[]$ (an empty list)
$\#s$	$length\ [\![s]\!]_{Isab}$
$s.(n+1)$	$nth\ [\![s]\!]_{Isab}\ [\![n]\!]_{Isab},\ [\![s]\!]_{Isab}\ !\ [\![n]\!]_{Isab}$
$x^\frown y$	$[\![x]\!]_{Isab}\ @\ [\![y]\!]_{Isab}$
$x \sqsubseteq y$	$[\![x]\!]_{Isab} \leq [\![y]\!]_{Isab},\ x, y \in M^*$
$x \sqsubseteq y$	$False,\ x \in M^*,\ x \in M^{\underline{\omega}}$
dom.s	$finU\_dom\ [\![s]\!]_{Isab}$ (or $finU\_dom\_inat\ [\![s]\!]_{Isab}$)
rng.s	$finU\_range\ [\![s]\!]_{Isab}$
$x \in s\ (s \in M^*$ for some $M)$	$[\![x]\!]_{Isab}\ mem\ [\![s]\!]_{Isab}$
ft.s	$hd\ [\![s]\!]_{Isab}$
$M \circledS s$	$filter\ (\lambda\, y.y \in [\![M]\!]_{Isab})\ [\![s]\!]_{Isab}$
map(f, s)	$map\ [\![f]\!]_{Isab}\ [\![s]\!]_{Isab}$
$\propto.s$	$remdups\ [\![s]\!]_{Isab}$

Table 2.4.: Isabelle/HOL representation of the FOCUS operators on finite untimed streams

---

[18] In these tables do not represent operators with complicated type combinations: prefix of a stream $x \sqsubseteq y$, concatenation operator $x^\frown y$,

Focus operator, f	Isabelle/HOL representation, $[\![f]\!]_{Isab}$
$\#s$	∞
$s.(n+1)$	$[\![s]\!]_{Isab} \; [\![n]\!]_{Isab}$
$x \frown y$	$[\![x]\!]_{Isab}$, x is infinite
$x \frown y$	$fin\_inf\_append \; [\![x]\!]_{Isab} \; [\![y]\!]_{Isab} \; (x \in M^*, \; y \in M^\infty)$
$x \sqsubseteq y$	$(\forall \, i. \; [\![x]\!]_{Isab} \; i = [\![y]\!]_{Isab} \; i), \; x, y \in M^\infty$
$x \sqsubseteq y$	$False, \; x \in M^\infty, y \notin M^\infty$
dom.s	$infU\_dom \; [\![s]\!]_{Isab}$
rng.s	$infU\_range \; [\![s]\!]_{Isab}$
$M \, \circledS \, s$	$filter\_inf \; (\lambda \, y.y \in [\![M]\!]_{Isab}) \; [\![s]\!]_{Isab}$
map(f, s)	$[\![f]\!]_{Isab} \; \circ \; [\![s]\!]_{Isab}$
$\propto.s$	$inf\_remdups \; [\![s]\!]_{Isab}$

Table 2.5.: Isabelle/HOL representation of the Focus operators on infinite untimed streams

Remarks to Table 2.8:

✓ In Isabelle/HOL can be used only the mathematical symbols $<$ and $\leq$, the symbols $>$ and $\geq$ do not belong to the Isabelle/HOL syntax.

✓ The mathematical symbols $<$ and $\leq$ can be used in Isabelle/HOL only for two arguments, because they are relations which results a boolean value (*True* or *False*). This implies that the expressions like $A < B < C$ contains type error.

✓ F denotes here some Focus function or predicate. If the expression is used in a simple definition, without any nestings, brackets can be omitted.

✓ P denotes here the name of some Focus predicate.

✓ T denotes some Focus datatype.

FOCUS operator, f	Isabelle/HOL representation, $[\![f]\!]_{Isab}$
$\langle\rangle$	$[]$ (an empty list)
$\#s$	$\mathit{fin\_length}\ [\![s]\!]_{Isab}$
$s.n$	$\mathit{fin\_nth}\ [\![s]\!]_{Isab}\ [\![n]\!]_{Isab}$
$x\frown y$	$[\![x]\!]_{Isab}\ @\ [\![y]\!]_{Isab}$
$x\sqsubseteq y$	$[\![x]\!]_{Isab}\leq[\![y]\!]_{Isab},\ x,y\in M^{\underline{*}}$
$x\sqsubseteq y$	$\mathit{inf\_prefix}\ [\![x]\!]_{Isab}\ [\![y]\!]_{Isab},\ x\in M^{\underline{*}},\ x\in M^{\underline{\infty}}$
$x\sqsubseteq y$	$\mathit{False},\ x\in M^{\underline{*}},\ y\in M^{\omega}$
$s\downarrow_t$	$\mathit{fin\_truncate\_plus}\ [\![s]\!]_{Isab}\ [\![t]\!]_{Isab}\ (t\in\mathbb{N}_{\infty})$
$s\downarrow_t$	$\mathit{fin\_truncate}\ [\![s]\!]_{Isab}\ [\![t]\!]_{Isab}\ (t\in\mathbb{N})$
$\mathsf{rng}.s$	$\mathit{finT\_range}\ [\![s]\!]_{Isab}$
$\overline{s}$	$\mathit{fin\_make\_untimed}\ [\![s]\!]_{Isab}$
$\mathsf{tm}(s,n)$	$\mathit{fin\_tm}\ [\![s]\!]_{Isab}\ [\![n]\!]_{Isab}$
$M\circledS s$	$\mathit{finT\_filter}\ [\![M]\!]_{Isab}\ [\![s]\!]_{Isab}$
$\mathsf{map}(f,s)$	$\mathit{map}\ [\![f]\!]_{Isab}\ [\![s]\!]_{Isab}$
$\propto.s$	$\mathit{finT\_remdups}\ [\![s]\!]_{Isab}$
$\mathsf{Nti}(s)$	$\mathit{length}\ [\![s]\!]_{Isab}$
$\mathsf{ti}(s,t)$	$\mathit{ti}\ [\![s]\!]_{Isab}\ [\![t]\!]_{Isab}$
$s\uparrow_t$	$\mathit{drop}\ [\![s]\!]_{Isab}\ [\![t]\!]_{Isab}$
$\mathsf{merge}^{ti}(s,r)$	$\mathit{fin\_merge\_ti}\ [\![s]\!]_{Isab}\ [\![r]\!]_{Isab}$
$\mathsf{ti}^k(s,n)$	$\mathit{fin\_join\_ti}\ [\![s]\!]_{Isab}\ [\![n]\!]_{Isab}\ [\![k]\!]_{Isab}$
$\mathsf{msg}_n(s)$	$\mathit{fin\_msg}\ [\![n]\!]_{Isab}\ [\![s]\!]_{Isab}$
$s\curlyvee_k$	$\mathit{fin\_join\_time}\ [\![s]\!]_{Isab}\ [\![k]\!]_{Isab}$
$s\curlywedge_k$	$\mathit{fin\_split\_time}\ [\![s]\!]_{Isab}\ [\![k]\!]_{Isab}$
$\mathsf{ttl}(s)$	$\mathit{tl}\ [\![s]\!]_{Isab}$
$\mathsf{fti}^{fin}(s)$	$\mathit{fin\_find1nonemp}\ [\![s]\!]_{Isab}$
$\mathsf{ind}_{fti}^{fin}(s)$	$\mathit{fin\_find1nonemp\_index}\ [\![s]\!]_{Isab}$
$\mathsf{last}^{ti}(s,t)$	$\mathit{fin\_last\_ti}\ [\![s]\!]_{Isab}\ [\![t]\!]_{Isab}$

Table 2.6.: Isabelle/HOL representation of the FOCUS operators on finite timed streams

FOCUS operator, f	Isabelle/HOL representation, $[\![f]\!]_{Isab}$
$s.n$	$inf\_nth\ [\![s]\!]_{Isab}\ [\![n]\!]_{Isab}$
$x \frown y$	$[\![x]\!]_{Isab},\ x$ is infinite
$x \frown y$	$fin\_inf\_append\ [\![x]\!]_{Isab}\ [\![y]\!]_{Isab}\ (x \in M^{*},\ y \in M^{\infty})$
$x \sqsubseteq y$	$(\forall\ i.\ [\![x]\!]_{Isab}\ i = [\![y]\!]_{Isab}\ i),\ x, y \in M^{\infty}$
$x \sqsubseteq y$	$False,\ x \in M^{\infty},\ y \notin M^{\infty}$
$s \downarrow_t\ (*)$	$inf\_truncate\_plus\ [\![s]\!]_{Isab}\ [\![t]\!]_{Isab}\ (t \in \mathbb{N}_{\infty})$
$s \downarrow_t\ (*)$	$inf\_truncate\ [\![s]\!]_{Isab}\ [\![t]\!]_{Isab}\ (t \in \mathbb{N})$
dom.s	$inf\_dom\ x$
rng.s	$infT\_range\ x$
ts(s)	$ts\ [\![s]\!]_{Isab}$
$\overline{s}$	$make\_untimed\ (InfT\ [\![s]\!]_{Isab})\quad (**)$
tm(s, n)	$inf\_tm\ [\![s]\!]_{Isab}\ [\![n]\!]_{Isab}$
$M \circledS s$	$infT\_filter\ [\![M]\!]_{Isab}\ [\![s]\!]_{Isab}$
map(f, s)	$[\![f]\!]_{Isab} \circ [\![s]\!]_{Isab}$
$\propto.s$	$infT\_remdups\ [\![s]\!]_{Isab}$
ti(s,t)	$[\![s]\!]_{Isab}\ [\![t]\!]_{Isab}$
$s \uparrow_t$	$inf\_drop\ [\![s]\!]_{Isab}\ [\![t]\!]_{Isab}$
merge$^{ti}(s,r)$	$inf\_merge\_ti\ [\![s]\!]_{Isab}\ [\![r]\!]_{Isab}$
ti$^k(s,n)$	$join\_ti\ [\![s]\!]_{Isab}\ [\![n]\!]_{Isab}\ [\![k]\!]_{Isab}$
msg$_n(s)$	$msg\ [\![n]\!]_{Isab}\ [\![s]\!]_{Isab}$
$s \curlyvee_k$	$join\_time\ [\![s]\!]_{Isab}\ [\![k]\!]_{Isab}$
$s \curlywedge_k$	$split\_time\ [\![s]\!]_{Isab}\ [\![k]\!]_{Isab}$
ttl(s)	$inf\_tl\ [\![s]\!]_{Isab}$
fti$^{inf}(s)$	$inf\_find1nonemp\ [\![s]\!]_{Isab}$
ind$_{fti}^{inf}(s)$	$inf\_find1nonemp\_index\ [\![s]\!]_{Isab}$
last$^{ti}(s, t)$	$inf\_last\_ti\ [\![s]\!]_{Isab}\ [\![t]\!]_{Isab}$
disj$_S^{inf}(l)$	$inf\_disjS\ [\![l]\!]_{Isab}$
disj$^{inf}(l)$	$inf\_disj\ [\![l]\!]_{Isab}$

Table 2.7.: Isabelle/HOL representation of the FOCUS operators on infinite timed streams (Remarks: ($*$) be careful with stream types; ($**$) the resulting type is α *stream*.)

FOCUS expression E	Isabelle/HOL representation, $[\![E]\!]_{Isab}$
true	*True*
false	*False*
$A = B$	$[\![A]\!]_{Isab} = [\![B]\!]_{Isab}$
$A \neq B$	$[\![A]\!]_{Isab} \neq [\![B]\!]_{Isab}$
$A < B,\, B > A$	$[\![A]\!]_{Isab} < [\![B]\!]_{Isab}$
$A \leq B,\, B \geq A$	$[\![A]\!]_{Isab} \leq [\![B]\!]_{Isab}$
$A < B < C,\, C > B > A$	$[\![A]\!]_{Isab} < [\![B]\!]_{Isab} \wedge [\![B]\!]_{Isab} < [\![C]\!]_{Isab}$
$A \leq B \leq C,\, C \geq B \geq A$	$[\![A]\!]_{Isab} \leq [\![B]\!]_{Isab} \wedge [\![B]\!]_{Isab} \leq [\![C]\!]_{Isab}$
$\neg Q$	$\neg [\![Q]\!]_{Isab}$
$Q_1 \vee Q_2$	$[\![Q_1]\!]_{Isab} \vee [\![Q_1]\!]_{Isab}$
$Q_1 \wedge Q_2$	$[\![Q_1]\!]_{Isab} \wedge [\![Q_2]\!]_{Isab}$
$Q_1 \Rightarrow Q_2$	$[\![Q_1]\!]_{Isab} \longrightarrow [\![Q_2]\!]_{Isab}$
$Q_1 \Leftrightarrow Q_2$	$[\![Q_2]\!]_{Isab} = [\![Q_2]\!]_{Isab}$
x	x (x is a variable)
$\exists\, x : Q$	$\exists\, x.\ [\![Q]\!]_{Isab}$
$\exists\, x_1 \ldots x_n : Q$	$\exists\, x_1 \ldots x_n.\ [\![Q]\!]_{Isab}$
$\forall\, x : Q$	$\forall\, x.\ [\![Q]\!]_{Isab}$
if F then F_1 else F_2 fi	$(if\ [\![F]\!]_{Isab}\ then\ [\![F_1]\!]_{Isab}\ else\ [\![F_2]\!]_{Isab})$
let A in B	$let\ [\![A]\!]_{Isab}\ in\ ([\![B]\!]_{Isab})$
$F(x_1, \ldots, x_n)$	$(F\ [\![x_1]\!]_{Isab} \ldots [\![x_n]\!]_{Isab})$
$\forall\, x_1 \ldots x_n \in T : Q$	$\forall\, (x_1 {::} [\![T]\!]_{Isab}) \ldots (x_n {::} [\![T]\!]_{Isab}).\ [\![Q]\!]_{Isab}$
$\exists\, x_1 \ldots x_n \in T : Q$	$\exists\, (x_1 {::} [\![T]\!]_{Isab}) \ldots (x_n {::} [\![T]\!]_{Isab}).\ [\![Q]\!]_{Isab}$
$\exists\, x_1 \ldots x_n \in [1..m] : Q$	$\exists\, x_1 \ldots x_n.$ $x_1 < m \wedge \cdots \wedge x_n < m \wedge [\![Q]\!]_{Isab}$
$\forall\, x_1 \ldots x_n \in [1..m] : Q$	$\forall\, x_1 \ldots x_n.$ $x_1 < m \wedge \cdots \wedge x_n < m \longrightarrow [\![Q]\!]_{Isab}$
$\forall\, x_1 \ldots x_n,\ P(x_1, \ldots, x_n) : Q$	$\forall\, x_1 \ldots x_n. [\![P(x_1, \ldots, x_n)]\!]_{Isab} \longrightarrow [\![Q]\!]_{Isab}$
$\exists\, x_1 \ldots x_n,\ P(x_1, \ldots, x_n) : Q$	$\exists\, x_1 \ldots x_n. [\![P(x_1, \ldots, x_n)]\!]_{Isab} \wedge [\![Q]\!]_{Isab}$

Table 2.8.: Isabelle/HOL representation of the FOCUS expressions

For example, according to Table 2.8 the Focus expression

$$\forall\, x_1 \ldots x_n \in [1, \ldots, m],\ P(x_1, \ldots, x_n) : Q$$

where $P(x_1, \ldots, x_n)$ is some Focus predicate, will be translated into Isabelle by

$$\forall\, x_1 \ldots x_n . x_1 < m \wedge \cdots \wedge x_n < m \wedge [\![P(x_1, \ldots, x_n)]\!]_{Isab} \ \longrightarrow\ [\![Q]\!]_{Isab}$$

The **where** statement is in Focus of the form

$$F \text{ where } v_1, \ldots, v_n \text{ so that } G$$

where $v_1, \ldots, v_n$ are logical variables and F, G are formulas (the formula G defines the values of the variables $v_1, \ldots, v_n$). The semantics of this statement is defined in Focus by the formula

$$\exists\, v_1, \ldots, v_n : F \wedge G$$

The original Focus definition allows to have as v_i ($1 \leq i \leq n$) also a function name – the function is then specified by the formula G. To have more clear syntax, we recommend to restrict the **where** statement the cases where $v_1, \ldots, v_n$ do not represent recursive functions or predicates. Always, if in the specification some auxiliary recursive function or predicate is needed, we recommend to define this function or predicate externally, in the special Focus frame.

For the case of non-recursive predicates defined in the **where** statement, we can say that the **where** statement is dual to the **let** statement

$$\text{let } G \text{ in } F$$

For this case we have

$$[\![F \text{ where } v_1, \ldots, v_n \text{ so that } G]\!]_{Isab} = \text{let } [\![G]\!]_{Isab} \text{ in } ([\![F]\!]_{Isab})$$

A Focus specification of a function *Name* with the signature $M_1 \times \cdots \times M_n \to M_{n+1}$ will be given using the following Focus frame:

```
┌─ Name ─────────────────────────────────────────────────
│
│   M_1 × ··· × M_n → M_{n+1}
│ ───────────────────────
│
│   Body
│
└────────────────────────────────────────────────────────
```

A Focus specification of a predicate *Name* with the signature $M_1 \times \cdots \times M_n \to$ Bool will be given in a similar way:

Name ───

$x_1 \in M_1; \quad \ldots; \quad x_n \in M_n$

───────────

Body

Body of the function or predicate *Name* can be of two types

✓ recursive definition[19], and

✓ non-recursive definition, abbreviation.

⚠ **Remark:** The semantics of the FOCUS functions (predicates) types is defined as follows:

$$[\![M_1 \times \cdots \times M_n \rightarrow M_{n+1}]\!]_{Isab} \equiv [\![M_1]\!]_{Isab} \Rightarrow \ldots \Rightarrow [\![M_n]\!]_{Isab} \Rightarrow [\![M_{n+1}]\!]_{Isab}$$

Body of recursive defined function or predicate looks like

BaseCase

InductionCase

The Isabelle/HOL semantics of this function will be the following one (for the case of predicate $M_{n+1} = \mathbb{B}ool$):

consts
  $Name :: [\![M_1]\!]_{Isab} \Rightarrow \ldots \Rightarrow [\![M_n]\!]_{Isab} \Rightarrow [\![M_{n+1}]\!]_{Isab}$
primrec
  $[\![BaseCase]\!]_{Isab}$
  $[\![InductionCase]\!]_{Isab}$

⚠ **Remark:** If in the FOCUS specification of a function or of a predicate non-primitive recursion is used (e.g. the recursion steps are larger as the recursion definition of the datatype of the parameter choosen for recursion), the syntax will be more complicated:

consts
  $Name :: [\![M_1]\!]_{Isab} \times \cdots \times [\![M_n]\!]_{Isab} \Rightarrow [\![M_{n+1}]\!]_{Isab}$
recdef $Name \ measure \ TerminationFunction$
  $[\![BaseCases]\!]_{Isab}$
  $[\![InductionCases]\!]_{Isab}$

where *TerminationFunction* is some function over $[\![M_1]\!]_{Isab} \times \cdots \times [\![M_n]\!]_{Isab}$ which shows that the function (the predicate) *Name* terminates.

Body of a function *Name* defined by *Formula*, i.e. like

$Name(x_1, \ldots, x_n) = Formula$

Its Isabelle/HOL semantics will be

---

[19]We strongly recommend to use primitive recursion (like Isabelle/HOL *primrec*-definition, see Section 1.2) to have simpler proofs about function (or predicate) properties.

---

constdefs

$Name :: [\![M_1]\!]_{Isab} \Rightarrow \ldots \Rightarrow [\![M_n]\!]_{Isab} \Rightarrow [\![M_{n+1}]\!]_{Isab}$

$Name\ x_1 \ldots x_n \equiv [\![Formula]\!]_{Isab}$

Body of recursive defined predicate looks like

$Formula_1$

...

$Formula_m$

The Isabelle/HOL semantics of such a predicate will be the following one:

constdefs

$Name :: [\![M_1]\!]_{Isab} \Rightarrow \ldots \Rightarrow [\![M_n]\!]_{Isab} \Rightarrow \mathbb{B}ool$

$Name\ x_1 \ldots x_n \equiv$

$[\![Formula_1]\!]_{Isab}$

$\wedge \ldots$

$\wedge [\![Formula_m]\!]_{Isab}$

where $x_1, \ldots, x_n$ represent the predicate parameters used in formulas $Formula_1$, ..., $Formula_m$.

 Remark: M_i, $1 \le i \le n+1$ can be some datatype as well as a stream of some datatype D, timed or untimed, finite or infinite:

$$M_i = D \mid D^* \mid D^\infty \mid D^{\underline{*}} \mid D^{\underline{\infty}}$$

We not allow to use the cases $M_i = D^\omega$ and $M_i = D^{\underline{\omega}}$ (there is no such a restriction in original Focus), because they imply too complicated proofs in Isabelle/HOL.

Example 2.8:

The function *Unpack* converts a finite list of entities of type

type $Package = pack(package\_id \in \mathbb{N}, package\_data \in D)$

where D is some datatype with $[\![D]\!]_{Isab} = D$.

Focus specification of the function *Unpack* can be represented as follows:

_Unpack_____

$Package^* \rightarrow D^*$

$Unpack(\langle\rangle) = \langle\rangle$

$Unpack(x \mathbin{\&} xs) = package\_data(x) \mathbin{\&} Unpack(xs)$

Now we translate this function by the schema:

consts
    Unpack :: "$[\![Package]\!]_{Isab} \Rightarrow [\![D]\!]_{Isab}$"
primrec
    $[\![Unpack(\langle\rangle) = \langle\rangle]\!]_{Isab}$
    $[\![Unpack(x \& xs) = package\_data(x) \& Unpack(xs)]\!]_{Isab}$

where

$[\![Package]\!]_{Isab} \equiv$ **record** *Package* =
                     package_id :: *nat*
                     package_data :: *D*

$[\![Unpack(\langle\rangle) = \langle\rangle]\!]_{Isab}$
$\equiv Unpack\ [\![\langle\rangle]\!]_{Isab} = [\![\langle\rangle]\!]_{Isab}$
$\equiv Unpack\ [] = []$

$[\![Unpack(x \& xs) = package\_data(x) \& Unpack(xs)]\!]_{Isab}$
$\equiv Unpack\ [\![x \& xs]\!]_{Isab} = [\![package\_data(x) \& Unpack(xs)]\!]_{Isab}$
$\equiv Unpack\ ([\![x]\!]_{Isab}\ \#\ [\![xs]\!]_{Isab}) = [\![package\_data(x)]\!]_{Isab}\ \#\ [\![Unpack(xs)]\!]_{Isab}$
$\equiv Unpack\ (x\ \#\ xs) = (package\_data\ [\![x]\!]_{Isab})\ \#\ (Unpack\ [\![xs]\!]_{Isab})$
$\equiv Unpack\ (x\ \#\ xs) = (package\_data\ x)\ \#\ (Unpack\ xs)$

Thus, we get at the end of translation of the definition of the corresponding function in Isabelle/HOL:

consts
    Unpack :: *Package* $\Rightarrow$ *D*
primrec
    Unpack $[] = []$
    Unpack $(x\ \#\ xs) = (package\_data\ x)\ \#\ (Unpack\ xs)$

                                                                    □

2.15. Summary

In this chapter we have introduced the implementation of the formal specification framework FOCUS in Isabelle/HOL. The following questions have been discussed here: which of the streams representation approaches is more appropriate for the case of FOCUS specifications of embedded real-time systems, which kind of graphical specification techniques is especially appropriated in this case, as well as which kind of FOCUS constructions are not very well situated to the direct translation to Isabelle/HOL and how we can reformulate them without changing their semantics.

We have presented in this chapter the translation of the FOCUS specifications of embedded real-time systems into Isabelle/HOL for proving properties of these systems. As result we get the deep embedding of that part of the framework FOCUS, which is appropriate for specification of real-time systems, into Isabelle/HOL:

- ✓ representation of FOCUS datatypes in Isabelle/HOL,

- ✓ representation of FOCUS streams in Isabelle/HOL,

- ✓ representation of the FOCUS operators on streams (length of a stream, nth message of a stream, concatenation operator, prefix of a stream, truncation operator, domain and range of a stream, time stamp operator, and stuttering removal operator),

- ✓ specification semantics and techniques,

- ✓ representation of the FOCUS extras: encapsulated states (local variables, control states, oracles), sheaf of channels and specification replication.

We have specified in this chapter the syntax extensions for FOCUS for the argumentation over time intervals: a special kind of tables tiTable, timed state transition diagrams and a number of new operators (the deep embedding includes these extensions), s.t. time interval operator, timed merge, timed truncation operator, limited number of messages per time interval, stuttering removal operator for timed streams, changing time granularity, deleting the first time interval, the last nonempty time interval until some time interval, and number of time intervals in a finite timed stream.

To prove correctness of the relations between the sets of input, output and local channels in Isabelle/HOL automatically, we have provided here a number of Isabelle/HOL theories and the corresponding proof schemata. These proof schemata for the correctness properties are standard and can be used automatically. If the proof fails, the specification of corresponding set is incorrect and must be changed.

3. Specification and Verification Methodology

This chapter presents the ideas of the specification and verification methodology for the approach "FOCUS on Isabelle", as well as the ideas of the so-called refinement-based verification. A number of case studies (see Chapter 4) has shown that we can influence on the Isabelle/HOL proofs complexity and reusability for a translated FOCUS specification also on the specification phase and not only on the verification phase.[1] Therefore, we want to concentrate on the possible specification modifications (reformulations), which can simplify the Isabelle/HOL proofs for a translated FOCUS specification. The proof strategy is also important, but plays at this point only a secondary role. Like in natural language a question contains a part of its answer, the proof of component or system properties is partially contained in the corresponding specification (see also Section 3.4). To simplify the work at the phase, when properties are verified, ones need to specify a component or system in the methodological way, taking into account the special properties of embedded real-time systems as well as the properties of Isabelle/HOL. Thus, we see specification and verification as strong bound phases.

The main questions discussed in this chapter are:

✓ In which way the refinement relation between specifications can be proved in Isabelle/HOL?

✓ Which parts in the specifications must be done more carefully and what is important to check to prevent mistakes, that are as experience more usual than another ones?

✓ Can we use the same (or similar) translation technics also for JANUS? Which kinds of specifications and operators of refinement steps are different from the corresponding ones in FOCUS?

3.1. Refinement

We can see any proof about a system as the proof that a (more concrete) system is a refinement of a (more abstract) one. The case when one needs to prove a single property of a system specification S can also be seen as a refinement relation: this property can be defined as a FOCUS specification S' itself and then one needs just show that the system specification S is a refinement of the specification S'.

---

[1]Here we mean proofs of system properties and not not proofs of correct relations between sets of channels, which were discussed in Sections 2.11.4 and 2.13.3.

In FOCUS (and in JANUS also) we can have a general specification S_0 of a system that corresponds to the formalization of system requirements. Therefore, in order to show that our concrete specification S_n that we get after n refinement steps fulfills the system requirements, we only need to show that the specification S_n is a refinement ([BS01, Bro97, Bro97]) of the specification S_0.

In this context, it is an important point what exactly a developer means by "refinement" on each refinement step (a behavioral refinement, an interface refinement, or a conditional refinement, changing time granularity etc.) and which specification semantics (FOCUS or JANUS, and respectively component or service) is used.

In this section we discuss first of all the representation of the refinement layers of a specification group and the general ways of their representation in Isabelle/HOL. After that we discuss the representation in Isabelle/HOL of different kinds of refinement – behavioral, interface, and conditional refinement.

3.1.1. Refinement Layers of a Specification Group

Figure 3.1 represents the hierarchy in a specification group S in general. It has m refinement layers:

✓ specification S^1 is a refinement specification of S,

✓ S^2 is a refinement specification of S^1,

✓ ...,

✓ S^j is a composition of specifications $S^j_1, \ldots, S^j_n$ (where for the specifications $S^j_1, \ldots, S^j_n$ the refinement layer j is the most abstract one) that builds a refinement of S^{j-1}.

✓ ...,

✓ S^m_i, $1 \leq i \leq n$ is a composition of specifications $S^m_{1,k1}, \ldots, S^m_{n,kn}$ that builds a refinement of S^{m-1}, where the specifications $S^m_{1,k1}, \ldots, S^m_{n,kn}$ are elementary ones.

The number N of all specification in the group is larger or equal[2] to the number of layers.

3.1.2. Behavioral Refinement

According to the definition of behavioral refinement (Definition 1.4), in order to show that the more concrete specification S_2 (e.g. a specification of a system architecture) fulfills the more abstract S_1 (e.g. system requirements), we only need to show

$$[\![S_2]\!] \Rightarrow [\![S_1]\!] \tag{3.1}$$

---

[2] Equality is possible only in the case, when we do not have any compositional specification in the group.

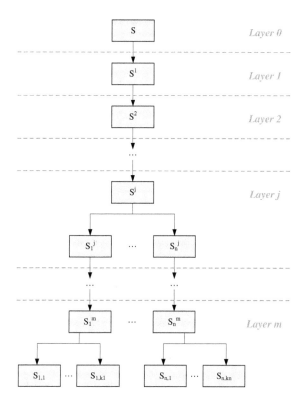

Figure 3.1.: Refinement Layers of a Specification Group S

In Isabelle it means to prove that the formula that corresponds to $[\![S_2]\!]$ implies the formula that corresponds to $[\![S_1]\!]$.

Definitions 1.1 and 1.3 (semantics of an elementary an a composite specification respectively) imply that the semantics of any FOCUS specification S can be represented by

$$i_S \in I_S^\infty \ \wedge \ o_S \in O_S^\infty \ \wedge \ B_S \qquad (3.2)$$

where i_S and o_S denote sets of input and output channel identifiers, I_S and O_S denote their corresponding types, and B_S is a logic formula. In the case of composite specification consisting of n subspecifications $S_1, \ldots, S_n$:

✓ The formula $[\![B_S]\!]_{Isab}$ is equal to formula

$$\exists \, l_S \in L_S^\infty : \bigwedge_{j=1}^{n} [\![B_{S_j}]\!]_{Isab}$$

where l_S denotes a list of local channel identifiers and L_S denotes their corresponding types, and B_{S_j} denotes the logic formula describing the body of the specification S_j.

✓ The lists of input and output channel identifiers, i_S and o_S, is a concatenation of corresponding lists of specifications $S_1, \ldots, S_n$, except those that are used as local channels (belong to l_S).

Together with Equation 1.5, this implies that the formal definition of the behavioral refinement [BS01] allows that the refined specification may have more input and output channels in addition to the input and output channels of the abstract specification. Let the specification S_2 be a behavioral refinement of the specification S_1 in the meaning of Definition 1.4: $S_1 \rightsquigarrow S_2$. According to the Equation 3.2 the semantics of these specification can be represented as follows:

$$[\![S_1]\!] = i_{S_1} \in I_{S_1}^{\infty} \ \wedge \ o_{S_1} \in O_{S_1}^{\infty} \ \wedge \ B_{S_1}$$
$$[\![S_2]\!] = i_{S_2} \in I_{S_2}^{\infty} \ \wedge \ o_{S_2} \in O_{S_2}^{\infty} \ \wedge \ B_{S_2}$$

Then, according to the Equation 1.5 we can conclude the following:

$$(S_1 \rightsquigarrow S_2) \iff$$
$$([\![S_2]\!] \Rightarrow [\![S_1]\!]) \Leftrightarrow$$
$$(i_{S_2} \in I_{S_2}^{\infty} \ \wedge \ o_{S_2} \in O_{S_2}^{\infty} \ \wedge \ B_{S_2}) \Rightarrow (i_{S_1} \in I_{S_1}^{\infty} \ \wedge \ o_{S_1} \in O_{S_1}^{\infty} \ \wedge \ B_{S_1})$$
$$\Leftrightarrow$$
$$(i_{S_2} \in I_{S_2}^{\infty} \Rightarrow i_{S_1} \in I_{S_1}^{\infty}) \ \wedge \ (o_{S_2} \in O_{S_2}^{\infty} \Rightarrow o_{S_1} \in O_{S_1}^{\infty}) \ \wedge \ (B_{S_2} \Rightarrow B_{S_1})$$

The conjunct

$$i_{S_2} \in I_{S_2}^{\infty} \Rightarrow i_{S_1} \in I_{S_1}^{\infty}$$

means that the set of input channels of the specification S_1 is a subset of the set of input channels of the specification S_2: $i_{S_2} \subseteq i_{S_1}$.

On the other hand the conjunct

$$o_{S_2} \in O_{S_2}^{\infty} \Rightarrow o_{S_1} \in O_{S_1}^{\infty}$$

means that the set of output channels of the specification S_1 is a subset of the set of output channels of the specification S_2: $o_{S_2} \subseteq o_{S_1}$.

For the cases when the more strict version of behavioral refinement is needed, where both specifications (an abstract one and a refined one) must have exactly the same syntactic interface, we introduce a new definition of the behavioral refinement – the *strict behavioral refinement*.

Definition: 3.2

A specification S_2 is called a *strict behavioral refinement* $(S_1 \rightsquigarrow S_2)$ of a specification S_1 if

✓ they have exactly the same syntactic interface and

✓ any I/O history of S_2 is also an I/O history of S_1.

We define the relation $\rightsquigarrow$ of strict behavioral refinement by equivalence

$$(S_1 \rightsquigarrow S_2)$$
$$\Leftrightarrow \tag{3.3}$$
$$((i_{S_1} \in I_{S_1}^{\infty} = i_{S_2} \in I_{S_2}^{\infty}) \wedge (o_{S_1} \in O_{S_1}^{\infty} = o_{S_2} \in O_{S_2}^{\infty}) \wedge (B_{S_2} \Rightarrow B_{S_1}))$$

where i_{S_1} and i_{S_2} (o_{S_1} and o_{S_2}) denote lists of input (output) channel identifiers of the specifications S_1 and S_2 respectively, I_{S_1}, I_{S_2}, O_{S_1} and O_{S_2} denote their corresponding types, B_{S_1} and B_{S_2} are logic formulas in terms of the Equation 3.2.

□

⚠ **Remark:** The definition of refinement does not exclude that the set of I/O histories of S_2 is empty. It means $[\![S_2]\!]$ is false and the refinement relation is true. This can happen if the specification S_2 is inconsistent (see Section 3.2).

3.1.3. Interface Refinement

Interface refinement (see Section 1.3.6 and Figure 1.3) can be interpreted as behavioral refinement modulo two representation specifications D and U, therefore we can use here the same ideas like in Section 3.1.2:

$$[\![S_1 \overset{(D,U)}{\rightsquigarrow} S_2]\!]_{Isab} \equiv [\![D \succ S_2 \succ U]\!]_{Isab} \to [\![S_1]\!]_{Isab}$$

3.1.4. Conditional Refinement

Conditional refinement (see Section 1.3.6 and Figure 1.4) is generalisation of interface refinement (which is again a generalisation of behavioral refinement), therefore we can again use here the same ideas like in Section 3.1.2:

$$[\![S_1 \overset{(D,U)}{\rightsquigarrow}_C S_2]\!]_{Isab} \equiv [\![C]\!]_{Isab} \wedge [\![D \succ S_2 \succ U]\!]_{Isab} \to [\![S_1]\!]_{Isab}$$

For the case of conditional *behavioral* refinement the relation is rather simpler:

$$[\![S_1 \rightsquigarrow_C S_2]\!]_{Isab} \equiv [\![C]\!]_{Isab} \wedge [\![S_2]\!]_{Isab} \to [\![S_1]\!]_{Isab}$$

In some cases we also need to extend the specification of a component (of a system) by some new properties. For example, we have to specify a system S_1 without taking into account possible errors in the behavior and want to represent this kind of behavior in the next step as an extended specification S_2 which guarantee part G_2 will be looks like

$$G_2 = \text{if } X \text{ then } G_1 \text{ else } G_1'$$

where X is some case-variable (in most cases an oracle), that indicates for each time interval whether an error occur does occur, G_1 denotes the guarantee part of the specification S_1 and G_1' denotes the extension of the guarantee part to represent errors in the behavior. This extensions of the specification does not fulfill the refinement definition, but we can prove that the extended specification S_2 is a refinement of the specification S_1', where S_1' is exactly the specification S_1 with the extra assumption about error-free behavior. This means that we can verify that S_2 has the same behavior for the an error-free case as S_1.

Example 3.9:

Let A_1 be the assumption part of a specification S_1 and let A_2 be the assumption part of a specification S_2, assuming $A_1 = A_2$. Let G_1 be the guarantee part of a specification S_1 and let G_2 be the guarantee part of a specification S_2, where $G_2 = $ if X then G_1 else G_1', i.e. $G_2 = (X \rightarrow G_1) \wedge (\neg X \rightarrow G_1')$:

$$S_1 = A_1 \rightarrow G_1 \quad \text{and} \quad S_2 = A_2 \rightarrow G_2.$$

We need to show that $[\![S_2]\!] \Rightarrow [\![S_1]\!]$.

1. Let $A_1 = A_2 = $ false, then
   $(\text{false} \rightarrow G_2) \rightarrow (\text{false} \rightarrow G_1)$
   true $\rightarrow$ true
   true

2. Let $A_1 = A_2 = $ true, then
   $(\text{true} \rightarrow G_2) \rightarrow (\text{true} \rightarrow G_1)$
   $G_2 \rightarrow G_1)$
   $(X \rightarrow G_1) \wedge (\neg X \rightarrow G_1') \rightarrow G_1$
   This formula does not hold for the case $X = $ false.

We cannot prove the property $(A_2 \rightarrow G_2) \rightarrow (A_1 \rightarrow G_1)$ without the additional assumption about X. Thus, we can only prove the property $(A_2 \rightarrow G_2) \rightarrow (A_1 \wedge X \rightarrow G_1)$.

□

3.2. Consistency of a Specification

The definition of a refinement relation does not exclude that the set of I/O histories of refined specification is empty. In this case the semantics of this specification is false (the specification is inconsistent) and the refinement relation is trivially true.

Let a specification S_2 be a refinement of a specification S_1. To show consistency is important only for the specification S_2, because the consistency of the abstract specification S_1 follows automatically from the proof of a refinement: if the specification S_1 is inconsistent (its semantics is false), the refinement relation holds only if the specification S_2 is also inconsistent.

There are two ways to prove consistency of the specification S_2:

(1) Inconsistency of the specification S_2 can be found out proving the refinement relation between S_1 and S_2: when the assumption part consisting of

the semantics of the specification S_2 becomes **false** without contradiction to some extra predicates that comes during the proof from the right part (semantics of S_1), the specification S_2 is inconsistent.

(2) The specification S_2 is consistent, if the set of its I/O histories is nonempty – if one shows that there exist some streams fulfilling the specification S_2, the specification S_2 is a consistent one. Thus, to show consistency in formal way it is enough to find an example of streams that fulfill the specification S_2. This way is more precise and formal, as the first one, but to find such an example is not trivial and for complicated, real systems it may be very difficult task.

3.3. Refinement-Based Verification

In the context of hardware and software systems, the definition of (formal) *verification* is "the act of proving or disproving the correctness of a system with respect to a certain formal specification or property, using formal methods of mathematics", where the definition of *validation* is "the quality control and testing tasks and techniques used to determine if a work product (either an application or one of its components) conforms to its specified requirements, including operational, quality, interface, and design constraints".

Thus, the verification means to proof properties of a system (more precisely, of a system specification) as some lemmas, where validation means to show that the refinement relation between specifications (more abstract one and more concrete one) holds. These concepts are very similar. Moreover, we can see verification of a system as a special case of validation: if the property to prove is presented as an abstract specification, it remains to validate the system specification with respect to these abstract specification, i.e. to show that the refinement relation holds. For example, the FOCUS specification *FlexRayArchitecture* (see Section 2.13.2) fulfills the property

$$\forall\, i \in [1..n] : \mathsf{msg}_1(get_i) \wedge \mathsf{msg}_1(store_i) \tag{3.4}$$

and

$$FrameTransmission(return_1, ..., return_n, store_1, ..., store_n, get_1, ..., get_n, c_1..., c_n) \tag{3.5}$$

under assumption that the properties

$$\forall\, i \in [1..n] : \mathsf{msg}_1(return_i)$$
$$DisjointSchedules(c_1, ..., c_n) \tag{3.6}$$
$$IdenticCycleLength(c_1, ..., c_n)$$

hold. Thus, the specification *FlexRayArchitecture* can be seen as

✓ a (behavioral) refinement of the specification *FlexRay'* that is combination of assumption from Equation 3.6 and guarantee from Equation 3.4.

✓   a (behavioral) refinement of the specification *FlexRay″* that is combination of assumption from Equation 3.6 and guarantee from Equation 3.5.

In most cases a system must be verified with respect not to a single property, but with respect to a number of properties. In this case it more sufficient to specify these properties as a single specification to exclude possible inconsistencies. Therefore, we verify that the specification *FlexRayArchitecture* is a (behavioral) refinement of the specification *FlexRay* that is combination of assumption from Equation 3.6 and guarantee from Equations 3.4 and 3.5. The semantics of this specification will be conjunction of the properties.

Let S be some specification of a system and let a specification L consist of properties $L_1, \ldots, L_n$ of this system. The corresponding refinement lemma looks like

$$[\![\, S \,]\!] \Rightarrow [\![\, L \,]\!]$$

Applying the definition of the Isabelle/HOL predicate which corresponds to $[\![\, L \,]\!]$ we get

$$[\![\, S \,]\!] \Rightarrow ([\![\, L_1 \,]\!] \wedge \cdots \wedge [\![\, L_n \,]\!])$$

Then we can split the verification goal into n subgoals, each of them can be proved as a separate lemma.

$$[\![\, S \,]\!] \Rightarrow [\![\, L_1 \,]\!]$$

$$\ldots$$

$$[\![\, S \,]\!] \Rightarrow [\![\, L_n \,]\!]$$

In some cases the requirements (properties) can be sorted to get a nested hierarchy. Assuming e.g. the following two requirements, namely L_1 and L_2, of some system S_1 that has an output channel y of type $\mathbb{N}$:

$$\forall\, t: \ \mathsf{ti}(y, t) \neq \langle \rangle \tag{3.7}$$

und

$$\forall\, t: \ \#\mathsf{ti}(y, t) = 2 \tag{3.8}$$

The second requirement, L_2, is a refinement of the first one, L_1. Therefore, if we show that the system S_1 fulfills the second requirement, we do not need to show that it fulfills the first requirement, but we need to show that the refinement relation between L_1 and L_2 holds, which in most cases is easier than to show that S_1 fulfills L_1.

Assuming a system S with corresponding list of requirements $L = [L_1, \ldots, L_n]$:

$$[\![\, S \,]\!] \Rightarrow [\![\, L \,]\!]$$

where

$$[\![\, L \,]\!] = [\![\, L_1 \,]\!] \wedge \cdots \wedge [\![\, L_n \,]\!]$$

For any new requirement R on the system S that we need to add to the list of its requirements L, $L \cup \{R\}$ (assuming R does not belong to the list of requirements) we can have the following cases.

(1) The system S has some requirement L_i that is less abstract than R:
$R \notin L \wedge \exists L_i \in L : L_i \Rightarrow R$.

We add R to the next level of abstraction L' (to the list with more abstract requirements, $[\![\ L \ ]\!] \Rightarrow [\![\ L' \ ]\!]$) using the same schema: $L' \cup \{R\}$, see Figure 3.2 (a).

(2) The list of requirements of the system S has a requirement that is more abstract than R:
$R \notin L \wedge \exists L_i \in L : R \Rightarrow L_i$.

We replace the requirement L_i in L by R, L_i will be added to the next level of abstraction L' (to the list with more abstract requirements), see Figure 3.2 (b).

If S does not fulfill R, then S must be changed according to the new list of requirements.

(3) The system S has no requirements that are in some relation (more/less abstract) to R (R opens some new "dimension" of S):
$R \notin L \wedge \forall L_i \in L : \neg(L_i \Rightarrow R) \wedge \neg(R \Rightarrow L_i)$.

For example, assuming the properties L_1 (see Equation 3.7) and L_2 (see Equation 3.8). The property R, defined by

$$\forall t : \ \mathsf{ft.ti}(y, t + 2) = 5,$$

does not imply L_1, because it is only about time intervals 2, 4, 6, 8 etc., and it also does not imply L_2, because it says nothing about the length of message list at the time intervals. Neither L_1 nor L_2 imply R, because they say nothing about the message values.

The R will be added to the list of requirements L, see Figure 3.2 (c). If S dos not fulfill R, then S must be changed according the new list of requirements.

The lists of requirements are specifications itself. Thus, we allude the specification hierarchy.

The legitimate question is, where we need to argue about such more abstract requirement like L_1 at all having more precise requirements like L_2, and why we cannot just remove them. The point is, that a number of system requirements comes out from the argumentation about interaction with another components or systems. Considering a system S_3 that consist of two subsystems: the system S_1 and some system S_2, and let the channel y be a local one for the system S_3, i.e. this output channel channel of S_1 will be an input channel for S_2. Thus, all assumptions of S_2 about this input channel y must be fulfilled by S_1 as its new requirements.

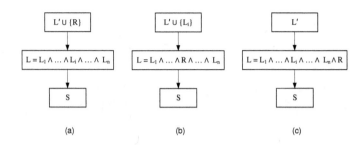

Figure 3.2.: Adding new requirement R to the list L of requirements of the specification S, $L \cup \{R\}$

3.4. Key Ideas of the Specification and Verification Methodology

To represent all the components of a specification group S in Isabelle/HOL we can use one of three definitions styles:

(1) All components are be specified as a *joined* Isabelle/HOL theory. In this case, we get only one theory *S.thy* with all ("body") specifications (see Fig. 3.3). This definition style is appropriate for systems with simple component specifications.

(2) Every component is specified as a *single* (separate) Isabelle/HOL theory. We get n Isabelle/HOL theories, where n is the number of all FOCUS component specifications in the specification group (see Fig. 3.4). This style is appropriate for systems with complicated component specifications and a small number of abstraction layers.

(3) All components *of the same abstraction* layer are specified as a *joined* Isabelle/HOL theory. We get m Isabelle/HOL theories, s.t. $m \leq n$, where m is the number of abstraction layers and n is the number of all FOCUS component specifications in the specification group (see Fig. 3.5). This style is appropriate for systems with complicated component specifications and a large number of abstraction layers.

The gray marked theories on Fig. 3.3-3.3 are predefined ones (see Sections 2.11.4 and 2.13.3, as well as Appendix A), the green marked theories are user-defined ones – they need to be defined doing the specification and verification of the group S.

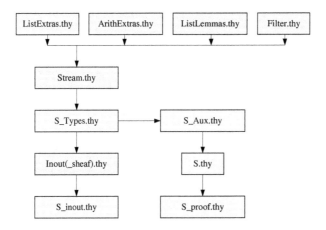

Figure 3.3.: Specification Group S: Representation of the whole group by a joined Isabelle/HOL theory

In all these cases we need to take into account that Isabelle/HOL representations[3] of all subcomponents $C_1, \ldots, C_n$ of the component C need to be defined

✓ for the cases (1) and (2): "above" the definition (or declaration) of the component C in the same theory as C.

✓ for the case (2) – in the theories on which is based the Isabelle/HOL theory for the component C:

$$C = \text{Main} + C_1 + \cdots + C_n$$

Verification of a large system S can be done in top-down as well as in bottom-up manner. Starting this process, we write the lemma S_L0 which says that the system architecture specification S fulfills its requirements specification $SReq$. The goal of this (main) lemma will be split into n subgoals, if the specification $SReq$ consists of n requirements. To make the prove of the main lemma readable, the lemmas S_L, ..., S_Ln are defined. Proving these lemmas we can find out, which properties of the subcomponents (or of composition of the subcomponents) are needed to solve the proof goals. The needed properties of a (single) subcomponent must either belong to the requirements specification of this subcomponent or are derivable from this specification (or from the architecture specification of this subcomponent). Thus, if the requirement specification is already defined, we deal with it as recommended in Section 3.3, otherwise we just define this specification to be conjunction of these properties.

---

[3] We speak about the same abstraction layer.

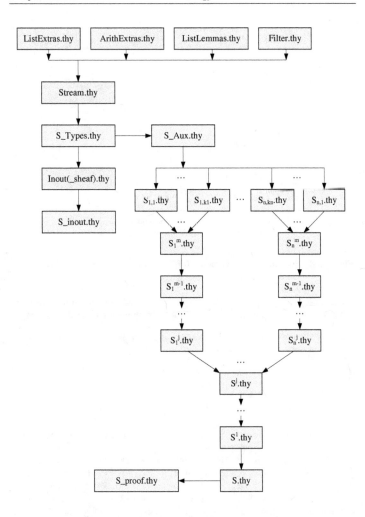

Figure 3.4.: Specification Group S: Representation of every component by a *single* (separate) Isabelle/HOL theory

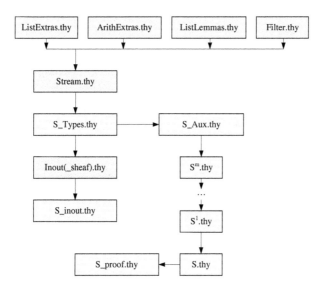

Figure 3.5.: Specification Group S: Representation of all components of the same abstraction layer by a *joined* Isabelle/HOL theory

Assuming we have composite specifications $A, \ldots, Z$, their subcomponents A_1, ..., A_m, ..., $Z_1, \ldots, Z_n$ and the corresponding requirements specifications $AReq, \ldots, ZReq$. If the refinement relations $AReq \rightsquigarrow A$, ..., $ZReq \rightsquigarrow Z$ have been proved, these requirements can be used to prove some properties some specification S, which is composed of $A, \ldots, Z$ – in most cases it is easier to prove the main lemma using these requirements specification, than to use the architecture specifications directly (see Section 4.3.10 for examples).

A main part of proofs of component (systems) properties in Isabelle/HOL could be done by functions and predicates definitions (defined by **constdefs**), induction (for recursive defined functions and predicates), Case-Tactics (if we have in the definition *case* or *if_then_else* parts), Isabelle/HOL automatic proof strategies, rules of natural induction, auxiliary arithmetics lemmas and lemmas about properties of system components.

The parts of proof, when we need to apply rules of natural deduction, as well as to define auxiliary arithmetics lemmas or lemmas about properties of system components, are really interactive and need efforts, but the cases of applying functions and predicates definitions, induction, Case-Tactics and simplifier can be solved schematically.

⚠ **Remark:** Isabelle/HOL is sensitive to the format in which the natural number are written, because the digit representation of a number corresponds to polymorphic numeral, i.e. such a number can also be of the integer type. Thus, it is always better to write its type explicit like $(0 :: nat)$, $(45 :: nat)$ etc. But even this can not prevent the following problem: the goal seems to be trivial to solve, but this does not happen, because the terms, which have equal semantics of a natural number, have different syntax. For example, if we have in the assumption part *(Suc (Suc (Suc (t + m))))* and in the goal *(t + 3 + m)* this goal cannot be solved directly. In this case the predefined *nat_number* lemma[4] must be used.

For the case we have a number of parenthesis, which lead to the different representation syntax of the natural number in the assumption and in the guarantee parts, e.g. $4 + (t + (k + d))$ and $4 + (d + (t + k))$, we need to define a subgoal that says *4 + (t + (k + d)) = 4 + (d + (t + k))* (such kind of subgoals can be solved trivially - by *arith*, *simp* or *auto*), and after that the proof of the goal is also trivial (see Section 4.3.10 for examples).

3.5. Specification Hints and Mistakes

A building of a formal specification and its refinements can also be error-prone. There is a number of mistakes classes that are as experience more usual than another ones. In this section we will discuss these classes to show which parts in the specifications must be done more carefully and what is important to check.

The following properties must be taken into account:

✓ System S is described by two FOCUS specifications S_1 (abstract one, e.g. requirement specification) and S_2 (more detailed one, e.g. architecture specification). If some assumptions in the definition of S_1 have been lost, we can find it out proving the refinement relation $S_1 \rightsquigarrow S_2$ ($[\![S_2]\!] \Rightarrow [\![S_1]\!]$). "Lost assumption" means here such an assumption, that cannot be derived from the definition of S_2 or, if S_2 is a composite specification, from the definition(s) of some subspecification(s) of S_2, but, conversely, in used in this definition (these definitions).

This rule does not work in the opposite direction.

✓ In Isabelle/HOL all functions and predicates used in the definition of some function or predicate f must be defined previous to the definition of f.

✓ Using the operator **ti** and the **tiTable**, as well as TSTDs to describe a component (a system) we can more easy prove the property of the component (the system), as arguing about the whole streams or single messages of the the streams. Arguing about time intervals, we need to

---

[4] The lemma *nat_number* converts the representations of all natural numbers to the *Suc*-form.

keep in mind, that a time interval is not a single message, but a list of messages (possibly containing a single message or empty).

✓ To avoid complicated specifications and proofs in Isabelle/HOL, additional local states instead of mutual recursive functions (see Section 2.9) must be used.

✓ Clear separation of the datatypes and the operation over datatypes leads to faster translation. In some cases type-mix can also lead to problems with type checker (see Section 4.1.1 for a concrete example).

✓ Combination of timed and time-synchronous streams also leads to problems with type checking. To avoid such problems, a time-synchronous stream must be represented as a *timed* stream for which a predicate *ts* holds. If we already have a completed FOCUS specification with the frame label "time-synchronous" or some input and/or output streams of a completed FOCUS specification with the frame label "timed" are time-synchronous, the translation schema from Section 2.10 must be used.

✓ "Time-synchronous" is more strict property of some stream s as $\mathsf{msg}_1(s)$.

✓ Specification using the operator "make stream untimed" leads to more complicated proofs, because many cases must be covered, e.g. if a stream contains infinitely many messages, the untimed version of this stream will be also infinite, otherwise the untimed version will be finite.

✓ It is simpler to use natural numbers instead of integer of rational numbers, because the argumentation over the set *nat* is much more straightforward.

✓ If a component receives some messages via some channel not regular and must forward every time interval the last of them, the following specification strategy is recommended: If it must be described how the component *processes* its input streams, it is better to use the representation with a local variable (this local variable models a buffer, where the last message will be stored). But in the case we describe a *requirement* on this component, it is better to use the operator last^{ti} (see Section 2.5 for the definition of the operator and Section 4.3.5 for an example).

✓ Avoid to use the filter operator on infinite *untimed* streams. The methodology "FOCUS on Isabelle" is mostly oriented on the timed streams, and, therefore, this application can be only the case for oracles. Thus, we can always reformulate the definition of an oracle, e.g.

$$\#\{true\} \, \textcircled{S} \, r = \infty \ \wedge \ \#\{false\} \, \textcircled{S} \, r = \infty$$

$$\equiv$$

$$\forall \, t : \ (r.t = true \Rightarrow \exists \, i.t < i \wedge r.i = false) \wedge$$
$$(r.t = false \Rightarrow \exists \, j.t < j \wedge r.j = true)$$

or

$$\{true\} \; \circledS \; p \neq \langle \rangle$$

$$\equiv$$

$$\exists \, i : \; r.i = true$$

✓ Do not name any stream o, because the letter o is reserved in Isabelle/HOL for the name of the composition operator.

✓ If the definition of some new type is not necessary, it is better to use predefined types ($\mathbb{B}$ool, $\mathbb{N}$, etc.).

3.6. Complementary Approach: JANUS on Isabelle

As mentioned in Section 1.4, the results of "FOCUS on Isabelle" can be also extended to a complementary approach, "JANUS on Isabelle", that presents a coupling of a JANUS with Isabelle/HOL.

The translation schema for functions and predicates will be exactly the same as for FOCUS (see Section 2.14). Isabelle/HOL specification of relation between sets of channels, as well as sheaves of channels, i.e. for both cases – with and without component (specification) replication, – an proof of their correctness (correctness of the syntactic interfaces) will be done exactly in the same way as presented in Sections 2.11.4 and 2.13.3 for the FOCUS specification.

Thus, we will have only two small varieties:

✓ The semantics of an A/G specification will be defined differently from the semantics of such a specification in FOCUS (see Section 2.11.2, Equation 2.27). In the case of the A/G specification of an JANUS service its body B_S consists of two parts – the assumption part B_S^A and the guarantee part B_S^G. The semantics of such a specification in JANUS is logical conjunction (in comparison with implication for the case of a component)

$$[\![B_S^{AG}]\!]_{Isab} = [\![B_S^A]\!]_{Isab} \; \wedge \; [\![B_S^G]\!]_{Isab} \tag{3.9}$$

where for B_S^A and B_S^G hold the same rules as in FOCUS (see Section 2.11.2).

✓ Some JANUS operators has other syntax or names as in FOCUS, as well JANUS has some new operators. These operators can easily be defined in Isabelle/HOL. For example:

◇ The JANUS operator $s.k$, kth sequence of the stream s – the sequence of messages communicated at time t in the stream s, corresponds to the time interval operator (see Section 2.5.3):

$$\mathsf{ti}(s, n) \stackrel{\text{def}}{=} s.(n + 1)$$

◇ The JANUS operator $s \downarrow k$, prefix of the first k sequences in the timed stream s, corresponds to the timed truncation operator (see Section 2.5.2):

$$s \downarrow k = s \downarrow_k$$

◇ The Isabelle/HOL semantics of the JANUS operator $A \circledS s$, sub-stream of s with only the elements in the set A, corresponds to the Isabelle/HOL semantics of the FOCUS filtering operator, because we have to define it in such way, that the time ticks $\sqrt{}$ are ignored (see Section 2.4.10).

◇ The JANUS operator $s \uparrow k$, stream s without the first k sequences, corresponds to the FOCUS operator $s \uparrow_k$ (see Section 2.5.4).

3.7. Summary

In this chapter we have discussed the ideas of the specification and verification methodology for the approach "FOCUS on Isabelle", as well as the ideas of the so-called refinement-based verification. Specification and verification are seen here as strong bound phases to focus at proofs approaches and Isabelle/HOL influences also during specification. We have discussed, in which way the refinement relation between two FOCUS specifications can be proved in Isabelle/HOL, as well as how all these ideas can be extended for the JANUS approach.

We have also introduced here the definition of *strict* behavioral refinement as well as presented a number of specification mistakes classes that are as experience more usual than another ones and the corresponding hints how one can prevent them.

4. Case Studies

Case studies are very important for research in the development of formal methods. They help us to find problems that we can get using methodology and also the corresponding solution to these problems. In this chapter we present three case studies that cover different application areas and the different specification elements to show feasibility of the approach:

✓ Steam Boiler System (process control),

✓ FlexRay communication protocol (data transmission),

✓ Automotive-Gateway System (memory and processing components, data transmission).

4.1. Steam Boiler System

For the first case study we choose a process control system, namely, a steam boiler control system. It can be represent as a distributed system consisting of a number of communicating components and must fulfill real time requirements. This case study shows how we can deal with local variables (states) and in which way we can represent mutually recursive functions to avoid problems in proofs. The main idea of the steam boiler specification was taken from [BS01]. The steam boiler has a water tank, which contains a number of gallons of water, and a pump, which adds 10 gallons of water per time unit to its water tank, if the pump is on. At most 10 gallons of water are consumed per time unit by the steam production, if the pump is off. The steam boiler has a sensor that measures the water level.

As mentioned in Section 2.11.5, where the syntactic interface of the system was discussed, the specification group *SteamBoiler* consists of the following components: *ControlSystem* (general requirements specification), *ControlSystemArch* (system architecture), *SteamBoiler*, *Converter*, and *Controller*. Thus, the steam boiler system is relatively small, and all its components can be specified as a joined Isabelle/HOL theory (see Section 3.4). We present the following Isabelle/HOL theories for this system:

✓ *SteamBoiler_types.thy* – the datatype definitions (presented in Section 2.11.5),
✓ *SteamBoiler_inout.thy* – the specification of component interfaces (relations between the sets of input and output channels) and the corresponding correctness proofs (presented in Section 2.11.5),
✓ *SteamBoiler.thy* – the Isabelle/HOL specifications of the system components,

✓ *SteamBoiler_proof* – the proof of refinement relation between the requirements and the architecture specifications.

In this section we discuss first of all the datatypes used in the case study. Then the FOCUS specifications of the components and their translation into Isabelle/HOL using the schema from Section 2.14 will be presented. After that the proof of the behavioral refinement relation between the requirements specification and the architecture specification will be discussed. The resulting Isabelle/HOL theories *SteamBoiler.thy* and *SteamBoiler_proof.thy* are presented in Appendix B.

4.1.1. Datatypes

The original steam boiler specification [BS01] in FOCUS uses two additional, user-defined, datatypes:

type *Gallons* $= \{r \in \mathbb{R} \mid 0 \leq r \leq 1000\}$
type *Switch* $= \{-1, +1\}$

The type *Gallons* can be replaced simply by the type $\mathbb{N}$, because specification of the water level by a rational number as well as the restriction of upper bound 1000 does not play an important role in this specification (especially taking into account the specification of requirements). Thus, we will replace in the whole steam boiler specification [BS01] the type *Gallons* by the type $\mathbb{N}$.

Introduction of the datatype *Switch* in the specification [BS01] aims to combine a number with arithmetical operations ("+" and "−") over it, to have shorter and clearer representation. But the use this type leads to problems with type checking: it is neither a subtype of *Gallons* nor of $\mathbb{N}$. This type was used in the steam boiler specification to represent the state of steam boiler pump (on/off). We will replace the type *Switch* by the type $\mathbb{B}$it (1 for "+1", 0 for "−1"). Thus, instead of

$o.j + (x.j) * r$

where $x.j$ was of type *Switch* we will use the representation

if $r = 0$ then $o.j - r$ else $o.j + r$ fi

Because not only time-synchronous specifications are used here, but also timed ones, we need to represent all of them in the *timed* frame and make corresponding changes to avoid problems with type checking (see also Section 3.5).

To have explicit difference between $0, 1 : \mathbb{N}$ and $0, 1 : \mathbb{B}$it we define in Isabelle/HOL the type $\mathbb{B}$it as follows:

datatype *bit* = *Zero* | *One*

4.1.2. Requirement Specification

The specification *ControlSystem* describes the requirements for the steam boiler system: in each time interval the system outputs it current water level in gallons

---

and this level should always be between 200 and 800 gallons (the system works in the time-synchronous manner). The original FOCUS specification of the control system component [BS01] is the following one:

```
══ ControlSystem ════════════════════════ time-synchronous ══
   out    o : Gallons
─────────────────────────────────────────────────────────────
   ∀ j ∈ ℕ₊ : 200 ≤ o.j ≤ 800
```

We make a number of changes wrt. the original specification:

✓ Types *Gallons* is replaced by $\mathbb{N}$.

✓ The stream o is renamed to s (see Section 3.5).

✓ Frame label *time-synchronous* is replaced by *timed*. We add new guarantee that the output stream is a time-synchronous one. The argumentation over elements of the stream $s.t$ is replaced by the argumentation over the first (and unique) element of the corresponding timed stream: ft.ti(s, t), and it goes now over the set $\mathbb{N}$ instead of the set $\mathbb{N}_+$ (see Section 2.10).

As result we get the following FOCUS specification of the component *Control-System*:

```
══ ControlSystem ═══════════════════════════════════ timed ══
   out    s : ℕ
─────────────────────────────────────────────────────────────
   ts(s)
   ∀ j ∈ ℕ : 200 ≤ ft.ti(s, j) ≤ 800
```

Now we convert the FOCUS specification into the corresponding Isabelle/HOL predicate schematically:

$$[\![(s)]\!] := ControlSystem]\!]_{Isab}$$
$$\equiv$$

constdefs

$\qquad ControlSystem :: nat\ istream \Rightarrow bool$

$\qquad ControlSystem\ s \equiv [\![\mathsf{ts}(s)]\!]_{Isab} \wedge [\![\forall j \in \mathbb{N} : 200 \leq \mathsf{ft.ti}(s, j) \leq 800]\!]_{Isab}$

where

$[\![\mathsf{ts}(s)]\!]_{Isab} \wedge [\![\forall j \in \mathbb{N} : 200 \leq \mathsf{ft.ti}(s, j) \leq 800]\!]_{Isab}$

$\equiv$

$(ts\ s) \wedge (\forall\ (j::nat).\ 200 \leq [\![\mathsf{ft.ti}(s, j)]\!]_{Isab} \wedge [\![\mathsf{ft.ti}(s, j)]\!]_{Isab} \leq 800)$

$\equiv$

$(ts\ s) \wedge (\forall\ (j::nat).\ 200 \leq hd\ [\![\mathsf{ti}(s, j)]\!]_{Isab} \wedge hd\ [\![\mathsf{ti}(s, j)]\!]_{Isab} \leq 800)$

$\equiv$

$(ts\ s) \wedge (\forall\ (j::nat).\ 200 \leq hd\ (s\ j) \wedge hd\ (s\ j) \leq 800)$

Thus, we get the following Isabelle/HOL definition:

constdefs
> *ControlSystem :: nat istream ⇒ bool*
> *ControlSystem s*
> ≡
> *(ts s)* ∧ *(∀ (j::nat).* 200 ≤ *hd (ti s j)* ∧ *hd (ti s j)* ≤ 800*)*

4.1.3. Architecture Specification

The specification *ControlSystemArch* describes one possible architecture of the steam boiler system. The system consists of three components: a steam boiler, a converter, and a controller.

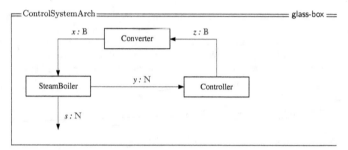

The corresponding FOCUS representation of this specification as plain text (constraint style):

> ══ ControlSystemArch ══════════════════════════════ timed ══
> out   *s* : N
> loc   *x, z* : Bit; *y* : N
> ───
> *(s, y) := SteamBoiler(x)*
> *(z) := Controller(y)*
> *(x) := Converter(z)*

Now we convert the FOCUS specification into Isabelle/HOL predicate *ControlSystemArch* according to the definition of semantics of composite specifications schematically (the Isabelle/HOL predicates *SteamBoiler*, *Converter* and *Controller* represent the Isabelle/HOL semantics of the FOCUS specifications of the components *SteamBoiler*, *Converter*, and *Controller* respectively).

$$\llbracket (s) := ControlSystemArch \rrbracket_{Isab}$$
$$\equiv$$

constdefs

$ControlSystemArch :: nat\ istream \Rightarrow bool$
$ControlSystemArch\ s$

$\equiv$

$[\![\ \exists\, x,\ z \in \mathbb{Bit}^{\infty} : \exists\, y \in \mathbb{N}^{\infty} :$
  $(s,\ y) := SteamBoiler(x)\ \wedge$
  $(z) := Controller(y)\ \wedge$
  $(x) := Converter(z)\ ]\!]_{Isab}$

where

$[\![\ \exists\, x,\ z \in \mathbb{Bit}^{\infty} : \exists\, y \in \mathbb{N}^{\infty} :$
  $(s,\ y) := SteamBoiler(x)\ \wedge\ (z) := Controller(y)\ \wedge\ (x) := Converter(z)]\!]_{Isab}$

$\equiv$

$\exists\ x\ z :: bit\ istream.\ \exists\ y :: nat\ istream.$
$(SteamBoiler\ x\ s\ y)\ \wedge\ (Controller\ y\ z)\ \wedge\ (Converter\ z\ x)$

⚠ **Remark:** The following Isabelle/HOL definition rule must be taken into account: all functions and predicates used in the definition of some function or predicate f must be defined previous to the definition of f. Thus, in the Isabelle/HOL theory *SteamBoiler.thy* the predicate *ControlSystemArch* must be defined after the predicates *SteamBoiler*, *Converter* and *Controller*. The predicate *ControlSystem* can be defined on any place before the proof part of the theory.

4.1.4. Steam Boiler Component

The specification *SteamBoiler* describes steam boiler component. The steam boiler works in time-synchronous manner: the current water level is controlled every time interval. The boiler has two output channels with equal streams $(y = s)$ and it fixes the initial water level to be 500 gallons. For every point of time the following must be true: if the pump is off, the boiler consumes at most 10 gallons of water, otherwise (the pump is on) at most 10 gallons of water will be added to its water tank.

The original FOCUS specification of this component [BS01] is the following one:

══ SteamBoiler ════════════════════════ time-synchronous ══
in $\quad$ $x : Switch$
out $\quad$ $y, o : Gallons$
$y = s\ \wedge\ s.1 = 500$ $\forall j \in \mathbb{N}_+ : \exists r \in Gallons : 0 < r \leq 10\ \wedge$ $\qquad\qquad\qquad s.(j+1) = o.j + (x.j) * r$

The changes wrt. the original specification:

✓ Types *Gallons* and *Switch* are replaced by $\mathbb{N}$ and $\mathbb{B}$it respectively.

✓ The stream o is renamed to s (see Section 3.5).

✓ Frame label *time-synchronous* is replaced by *timed*. We add new assumption that the input stream is time-synchronous, as well as new guarantee that the output streams are time-synchronous. The argumentation over elements of the stream $s.(t+1)$ is replaced by the argumentation over the first (and unique) element of the corresponding time stream: $\mathsf{ft.ti}(s, t)$, and it goes now over the set $\mathbb{N}$ instead of the set $\mathbb{N}_+$(see Section 2.10).

As result we get the following FOCUS specification:

```
┌─ SteamBoiler ══════════════════════════════════════════ timed ══
│  in     x : Bit
│  out    y, s : ℕ
├──────────────────────────────────────────────────────────────────
│  asm  ts(x)
├ ─ ─ ─ ─ ─ ─ ─ ─ ─ ─ ─ ─ ─ ─ ─ ─ ─ ─ ─ ─ ─ ─ ─ ─ ─ ─ ─ ─ ─ ─ ─
│  gar  ts(y)
│       ts(s)
│       y = s ∧ ft.ti(s, 0) = 500
│       ∀j ∈ ℕ : ∃r ∈ ℕ : 0 < r ≤ 10 ∧
│                ft.ti(s, j + 1) = if ft.ti(x, j) = 0 then ft.ti(s, j) − r else ft.ti(s, j) + r fi
└──────────────────────────────────────────────────────────────────
```

Now we represent the specification *SteamBoiler* in Isabelle/HOL.

$\llbracket (s, \ y) = SteamBoiler(x) \rrbracket_{Isab}$
$\equiv$

constdefs
    $SteamBoiler :: bit\ istream \Rightarrow nat\ istream \Rightarrow nat\ istream \Rightarrow bool$
    $SteamBoiler\ x\ s\ y$
    $\equiv$
    $(ts\ x)$
    $\longrightarrow$
    $((ts\ y) \wedge (ts\ s) \wedge (y = s) \wedge$
    $\llbracket \mathsf{ft.ti}(s, 0) = 500 \rrbracket_{Isab} \wedge$
    $\llbracket \forall j \in \mathbb{N} : \exists r \in \mathbb{N} : 0 < r \leq 10 \ \wedge$
      $\mathsf{ft.ti}(s, j + 1) =$
      $\text{if } \mathsf{ft.ti}(x, j) = 0 \text{ then } \mathsf{ft.ti}(s, j) − r \text{ else } \mathsf{ft.ti}(s, j) + r \text{ fi } \rrbracket_{Isab})$

where

$\llbracket \mathsf{ft.ti}(s, 0) = 500 \rrbracket_{Isab} = (hd\ (s\ 0) = (500::nat))$

and

$\llbracket \forall j \in \mathbb{N} : \exists r \in \mathbb{N} : 0 < r \leq 10 \ \wedge$
$\mathsf{ft.ti}(s, j + 1) =$

if ft.ti$(x, j) = 0$ then ft.ti$(s, j) - r$ else ft.ti$(s, j) + r$ fi $]\!]_{Isab}$

$\equiv$

$\forall$ *(j::nat).* $(\exists$ *(r::nat).*

(0::nat) $< r \wedge r \le$ *(10::nat)* $\wedge$

hd (s (Suc j)) =

   (if (hd (x j)) = Zero

   then   *(hd (s j)) - r*

   else   *(hd (s j)) + r))*

4.1.5. Converter Component

The specification *Converter* describes the converter component. It converts the asynchronous output produced by the controller to time-synchronous input for the steam boiler. Initially the pump is off, and at every later point of time (from receiving the first instruction from the controller) the output will be the last input from the controller $(\overline{z\!\downarrow}_t.(\#\overline{z\!\downarrow}_t))$. The original FOCUS specification of this component [BS01] is the following one:

⎓ Converter ⎓⎓⎓⎓⎓⎓⎓⎓⎓⎓⎓⎓⎓⎓⎓⎓⎓⎓⎓⎓ timed ⎓
in z : *Switch*
out x : *Switch*
ts(x) $\wedge$ $\forall t \in \mathbb{N} : \overline{x}.(t+1) =$ if $\overline{z\!\downarrow}_t = \langle\rangle$ then -1 else $\overline{z\!\downarrow}_t.(\#\overline{z\!\downarrow}_t)$ fi

The changes wrt. the original specification:

✓ Type *Gallons* is replaced by the type $\mathbb{B}it$.

✓ The expression $\overline{x}.(t+1)$ is replaced by ft.ti(x, t) (according to Equation 2.1) to have simpler proofs.

As result we get the following FOCUS specification:

⎓ Converter ⎓⎓⎓⎓⎓⎓⎓⎓⎓⎓⎓⎓⎓⎓⎓⎓⎓⎓⎓⎓ timed ⎓
in z : $\mathbb{B}it$
out x : $\mathbb{B}it$
ts(x) $\wedge$ $\forall t \in \mathbb{N} :$ ft.ti$(x, t) =$ if $\overline{z\!\downarrow}_t = \langle\rangle$ then 0 else $\overline{z\!\downarrow}_t.(\#\overline{z\!\downarrow}_t)$ fi

Now we convert the FOCUS specification into Isabelle/HOL predicate *Converter* schematically:

$[\!][(x) := Converter(z)]\!]_{Isab}$

$\equiv$

constdefs

    Converter :: bit istream $\Rightarrow$ bit istream $\Rightarrow$ bool

$$Converter\ z\ x$$
$$\equiv$$
$$(ts\ x)\ \wedge$$
$$[\!\![\forall\, t \in \mathbb{N} : \mathsf{ft.ti}(x, t) = \text{if}\ \ \overline{z{\downarrow}_t} = \langle\rangle\ \text{then}\ \ 0\ \text{else}\ \ \overline{z{\downarrow}_t}.(\#\overline{z{\downarrow}_t})\ \text{fi}\ ]\!\!]_{Isab}$$

where

$$[\!\![\forall\, t \in \mathbb{N} : \mathsf{ft.ti}(x, t) = \text{if}\ \overline{z{\downarrow}_t} = \langle\rangle\ \text{then}\ \ 0\ \text{else}\ \ \overline{z{\downarrow}_t}.(\#\overline{z{\downarrow}_t})\ \text{fi}\ ]\!\!]_{Isab}$$
$$\equiv$$
$$(\forall\ (t::nat).\ hd\ (x\ t) =$$
$$\quad (if\ [\!\![\overline{z{\downarrow}_t}]\!\!]_{Isab} = [\,]$$
$$\quad then\ Zero$$
$$\quad else\ [\!\![\overline{z{\downarrow}_t}.(\#\overline{z{\downarrow}_t})]\!\!]_{Isab}))$$
$$\equiv$$
$$(\forall\ (t::nat).\ hd\ (x\ t) =$$
$$\quad (if\ (fin\_make\_untimed\ (inf\_truncate\ z\ t) = [\,])$$
$$\quad then\ Zero$$
$$\quad else\ (fin\_make\_untimed\ (inf\_truncate\ z\ t))\ !$$
$$\qquad\qquad ((length\ (fin\_make\_untimed\ (inf\_truncate\ z\ t)))\ \text{-}1)\ ))$$

Here the streams $z{\downarrow}_t$ and $\overline{z{\downarrow}_t}$ are a finite timed one and a finite untimed one respectively, where the stream z is an infinite timed one.

4.1.6. Controller Component

The specification *Controller* describes the controller component. Contrary to the steam boiler the controller behaves in a purely asynchronous manner to keep the number of control signals small, it means it might not be desirable to switch the pump on and off more often than necessary. The controller is responsible for switching the steam boiler pump on and off.

If the pump is off $(\mathit{off}(\langle r\rangle \frown y))$: if the current water level is above 300 gallons the pump stays off $((\langle\sqrt{}\rangle\frown\mathrm{off}(y))$, otherwise the pump is started

$$\langle +1\rangle \frown \langle\sqrt{}\rangle \frown on(y)$$

and will run until the water level reaches 700 gallons.

If the pump is on $(\mathit{on}(\langle r\rangle \frown y))$: if the current water level is below 700 gallons the pump stays on $((\langle\sqrt{}\rangle\frown\mathrm{on}(y))$, otherwise the pump is turned off

$$\langle -1\rangle \frown \langle\sqrt{}\rangle \frown \mathit{off}(y)$$

and will be off until the water level reaches 300 gallons.

The original FOCUS specification of this component [BS01] is the following one:

Controller == **timed**

in $y : Gallons$

out $z : Switch$

$z = \langle -1 \rangle \frown \langle \sqrt{} \rangle \frown \mathit{off}\,(\overline{y})$
where off so that $\forall\, r \in Gallons;\;\; y \in Gallons^{\,\omega}$:
    $\mathit{off}\,(\langle r \rangle \frown y) =$ if $r > 300$ then $\langle \sqrt{} \rangle \frown \mathit{off}\,(y)$ else $\langle +1 \rangle \frown \langle \sqrt{} \rangle \frown \mathit{on}(y)$ fi
    $\mathit{on}(\langle r \rangle \frown y) \;=\;$ if $r < 700$ then $\langle \sqrt{} \rangle \frown \mathit{on}(y)$ else $\langle -1 \rangle \frown \langle \sqrt{} \rangle \frown \mathit{off}\,(y)$ fi

---

The specification of the controller is a *timed* one, but making a proof in Is-abelle/HOL that the architecture specification *ControlSystemArch* is a behavioral refinement of the requirement specification *ControlSystem*, we found out that to argue about its properties we need an assumption about the input stream y, that the stream y is a time-synchronous one: $\mathsf{ts}(y)$. In the steam boiler system specified by *ControlSystemArch* this assumption for the component *Controller* will be always *true* because the stream y is an output stream of the time-synchronous component *SteamBoiler*, but if we look at the component[1] *Controller* without taking into account its later combination[2] with the component *SteamBoiler*, then we may have some problems. More precisely, the specification *Controller* says, that when some input message comes, the increment by 1 of the global digital clock ($\sqrt{}$ in the output stream) happens. This means the following:

✓   If in some time interval there is no message in the input stream y, no increment of the clock happens.

✓   If in some time interval a number of messages comes, the clock will be incremented *for each* message.

Both cases lead to the wrong behavior of the component *Controller*, but for the steam boiler controller system it never happens, because the stream y is an output stream of the time-synchronous component *SteamBoiler*.

To argue about properties of controller and to have a possibility to reuse the component later, we extend the specification *Controller* by assumption predicate $\mathsf{ts}(y)$ to the specification *ControllerExt* (the datatypes *Gallons* and *Switch* are also replaced by $\mathbb{N}$ and $\mathbb{B}it$ respectively).

⚠ **Remark:** The specification *ControllerExt* is not a behavioral refinement of component *Controller*.

---

[1] This is true also for the case of service semantics.
[2] In most cases it is much more simpler to argue about a single component and its properties as about the whole system.

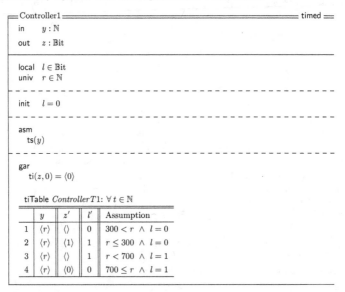

The specification above uses mutually recursive functions *on* and *off* to specify the local state of the component, more precisely, the state of the steam boiler pump. Because the translation of such functions to Isabelle/HOL and argumentation about them is very complicated (see Section 2.9), we need to reformulate the specification *ControllerExt* into a semantically equal specification that uses local states instead of mutually recursive functions. Therefore, we introduce a new local variable $l \in \mathbb{Bit}$ (0 corresponds to "off", 1 corresponds to "on") and set it initially to 0. Thus, we get FOCUS specification *Controller1* that is semantically equal to the FOCUS specification *ControllerExt*.

How it was shown in Section 2.6, we can rewrite the formula that represents semantics of the tiTable *ControllerT1* to the formula

```
if l = 0
then if 300 < r
      then ti(z, t + 1) = ⟨⟩ ∧ l' = 0
      else ti(z, t + 1) = ⟨1⟩ ∧ l' = 1 fi
else  if r < 700
      then ti(z, t + 1) = ⟨⟩ ∧ l' = 1
      else ti(z, t + 1) = ⟨0⟩ ∧ l' = 0 fi
fi
where r = ft.ti(y, t)
```

The proof that the formula above describes the behavior of the mutually recursive functions *on* and *off* from the specifications *Controller* and *ControllerExt* is straightforward.

We convert the FOCUS specification *Controller1* into Isabelle/HOL predicate *Controller* schematically:

$$[\![(z) := Controller1(y)]\!]_{Isab}$$
$$\equiv$$

constdefs

> *Controller* :: *nat istream* ⇒ *bit istream* ⇒ *bool*
> *Controller y z*
> $\equiv$
> *(ts y)* ⟶ *(∃ l. Controller_L y (fin_inf_append [Zero] l) r1 ... rk l1)*

where *Controller_L* is an auxiliary Isabelle/HOL predicate to represent the local variable l in the specification:

constdefs

> *Controller_L* ::
> > *nat istream* ⇒ *bit iustream* ⇒ *bit iustream* ⇒ *bit istream* ⇒ *bool*
> *Controller y lIn lOut z*
> $\equiv$
> $[\![\text{ti}(z, 0) = ⟨0⟩]\!]_{Isab}$ ∧
> $[\![\forall t \in \mathbb{N} :$
> if $l = 0$
> then if $300 < r$
> > then ti(z, t + 1) = ⟨⟩ ∧ l' = 0
> > else ti(z, t + 1) = ⟨1⟩ ∧ l' = 1 fi
> else if $r < 700$
> > then ti(z, t + 1) = ⟨⟩ ∧ l' = 1
> > else ti(z, t + 1) = ⟨0⟩ ∧ l' = 0 fi
> fi
> where $r = \text{ft.ti}(y, t)$ $]\!]_{Isab})$

where

$$\llbracket \mathsf{ti}(z,0) = \langle 0 \rangle \rrbracket_{Isab} = ((z\ 0) = [Zero])$$

$$\llbracket \forall\, t \in \mathbb{N} :$$
$$\quad \mathsf{if}\ l = 0$$
$$\quad \mathsf{then}\ \mathsf{if}\ 300 < r$$
$$\qquad\quad \mathsf{then}\ \mathsf{ti}(z, t+1) = \langle\rangle \wedge l' = 0$$
$$\qquad\quad \mathsf{else}\ \mathsf{ti}(z, t+1) = \langle 1 \rangle \wedge l' = 1\ \mathsf{fi}$$
$$\quad \mathsf{else}\ \ \mathsf{if}\ r < 700$$
$$\qquad\quad \mathsf{then}\ \mathsf{ti}(z, t+1) = \langle\rangle \wedge l' = 1$$
$$\qquad\quad \mathsf{else}\ \mathsf{ti}(z, t+1) = \langle 0 \rangle \wedge l' = 0\ \mathsf{fi}$$
$$\quad \mathsf{fi}$$
$$\quad \mathsf{where}\ \mathsf{ti}(y, t) = \langle r \rangle\ \rrbracket_{Isab})$$
$$\equiv$$
$$\forall\ (t{::}nat).$$
$$(\ if\ (lIn\ t) = Zero$$
$$\quad then\ \ (\ if\ 300 < hd\ (y\ t)$$
$$\qquad\quad then\ (z\ t) = []\ \wedge\ (lOut\ t) = Zero$$
$$\qquad\quad else\ (z\ t) = [One]\ \wedge\ (lOut\ t) = One$$
$$\qquad\quad)$$
$$\quad else\ \ (\ if\ hd\ (y\ t) < 700$$
$$\qquad\quad then\ (z\ t) = []\ \wedge\ (lOut\ t) = One$$
$$\qquad\quad else\ (z\ t) = [Zero]\ \wedge\ (lOut\ t) = Zero\ )\ )$$

4.1.7. Verification of the Steam Boiler System

To show that the specified system fulfills the requirements we need to show that the specification *ControlSystemArch* is a refinement of the specification *ControlSystem*. It follows from the definition of behavioral refinement that in order to verify that

$$ControlSystem \ \leadsto \ ControlSystemArch \tag{4.1}$$

it is enough to prove that

$$\llbracket ControlSystemArch \rrbracket \ \Rightarrow \ \llbracket ControlSystem \rrbracket \tag{4.2}$$

Therefore, we have to define and to prove a *lemma* (let us call it *L0_ControlSystem*), that says the specification *ControlSystemArch* is a refinement of the specification *ControlSystem*:

lemma *L0_ControlSystem:*
$\llbracket\ ControlSystemArch\ s \rrbracket \Longrightarrow ControlSystem\ s$

To prove this lemma we used first of all the definition of the predicate *Control-System* and then split the goal into three subgoals:

1. Stream s is a time-synchronous one:
   $ControlSystemArch\ s \Longrightarrow ts\ s$

2. The water level is always greater than 200 gallons:

   $\wedge$ j. ControlSystemArch s $\Longrightarrow$ 200 $\leq$ hd (s j)

3. The water level is always less than 800 gallons:

   $\wedge$ j. ControlSystemArch s $\Longrightarrow$ hd (s j) $\leq$ 800

The predicate *ts* holds for the stream *s* according to the definitions of the components *ControlSystemArch*, *SteamBoiler* and *Converter*. Thus, to solve the first goal we apply the definitions of the corresponding Isabelle/HOL predicates and the Isabelle/HOL automatic proof strategy.

To solve the second and the third goals we apply lemmas *L1_ControlSystem* and *L2_ControlSystem* respectively. These lemmas say that if the predicate *Control-SystemArch* holds for a stream *s*, then the first element of any time interval of this stream[3] is less or equal than 200 and greater or equal than 800 respectively.

lemma *L1_ControlSystem*:
ControlSystemArch s $\Longrightarrow$ (200::nat) $\leq$ hd (s i)

lemma *L2_ControlSystem*:
ControlSystemArch s $\Longrightarrow$ hd (s i) $\leq$ (800:: nat)

To prove the lemma *L1_ControlSystem* we used first of all the definition of the predicate *ControlSystemArch*. According to the definition of the component *SteamBoiler*, the value of *hd (s 0)* is specified as 500 and the value of *hd (s (Suc i))* is evaluated from the value of *hd (s i)*. Therefore, we need to apply the induction rule. The base case of induction can be proved by the definitions of the predicates *SteamBoiler* and *Converter*.

To prove the induction step we apply the definitions of all the components, the Isabelle/HOL automatic proof strategies, split the if_then_else-expressions, apply lemma *fin_make_untimed_nth_length* from the theory *stream.thy* (see Appendix A.1)[4] and lemma *last_nth_length* from the theory *ListExtras.thy* (see Appendix A.5)[5] , as well as lemmas about controllers properties, *L4_Controller* and *L3_Controller*).

The proof of the lemma *L2_ControlSystem* is analogous.

The lemma *L4_Controller* says that if for streams *s*, (fin_inf_append [Zero] l), l, and *z* the predicate *Controller_L* holds, then the stream *z*, which was truncated after *i*th time interval and after that made untimed, is a nonempty one:

---

[3]The stream *s* is defined in our system as time-synchronous one. Thus, every time interval of this stream consists of exactly one element.

[4]The lemma *fin_make_untimed_nth_length* says that if the *i*th time interval of the stream *z* consists of the element *a*, then the *j*th element of the stream *z*, which was truncated after *i*th time interval and after that made untimed, is equal *a* (where *j* is the length of the untimed truncated stream *z*).

[5]The lemma *last_nth_length* says that if the list *x* is nonempty one, then the *n*th its element, where *n* is the length of *x*, is the last element of the list.

lemma $L4\_Controller$:
$\llbracket$ $Controller\_L$ s $(fin\_inf\_append$ $[Zero]$ $l)$ l z $\rrbracket$
$\Longrightarrow$ $fin\_make\_untimed$ $(inf\_truncate$ z $i) \neq \llbracket\rrbracket$

To prove this lemma we used first of all the definition of the predicate $Controller\_L$. Then we add to the assumptions the fact, that the stream z, which was truncated after ith time interval and after that made untimed, is a nonempty one, and use the Isabelle/HOL automatic proof strategy.

The lemma $L3\_Controller$ says that if for streams s, $(fin\_inf\_append$ $[Zero]$ $l)$, l, and z the predicate $Controller\_L$ holds, then the last message of the stream z until the time i is equal to the ith message of the stream l:

$L3\_Controller$:
$\llbracket$ $Controller\_L$ y $(fin\_inf\_append$ $[Zero]$ $l)$ l $z\rrbracket$
$\Longrightarrow$
$last$ $(fin\_make\_untimed$ $(inf\_truncate$ z $t)) = l$ t

This lemma is proved using the *case* tactics over the value of the local variable l at time t ($Zero$, One), as well ass lemmas $L1\_Controller$ and $L2\_Controller$, which describe properties combinations of the controller component and the operators $fin\_make\_untimed$ and fin_inf_append (see Appendix A.1) and the Isabelle/HOL automatic proof strategy.

The whole proofs of the system properties are represented in the Appendix B.2.

4.1.8. Results of the Case Study

In this case study we have shown how we can deal with local variables (states) and in which way we can represent mutual recursive functions to avoid problems in proofs.

The FOCUS specification of the steam boiler system [BS01] was first of all extended according 3.5, after that the FOCUS specifications of all components of the system were translated schematically to Isabelle/HOL and the refinement relation between the requirement and the architecture specification of the system was proved. Proving this relation in Isabelle/HOL, we found out that to argue about properties of the *Controller* component of the system we need an additional (wrt. the original specification from [BS01]) assumption about the input stream y, that the stream y is a time-synchronous one.

The correctness of the input/output relations was also proved for all components of the system.

4.2. FlexRay Communication Protocol

In this section we present the Case Study on FlexRay, communication protocol for safety-critical real-time applications. This protocol has been developed by the FlexRay Consortium [Fle] for embedded systems in vehicles. The advantages of FlexRay over a CAN protocol (Controller Area Network, see [Rob91]), which is the most currently used protocol for such kind of systems, are deterministic real-time message transmission, fault tolerance, integrated functionality for clock synchronisation and higher bandwidth.

FlexRay contains a set of complex algorithms to provide the communication services. From the view of the software layers above FlexRay only a few of these properties become visible. The most important ones are static cyclic communication schedules and system-wide synchronous clocks. These provide a suitable platform for distributed control algorithms as used e.g. in drive-by-wire applications. The formalization described here is based on the "Protocol Specification 2.0"[Fle04].

The static message transmission model of FlexRay is based on *rounds*. FlexRay rounds consist of a constant number of time slices of the same length, so called *slots*. A node can broadcast its messages to other nodes at statically defined slots. At most one node can do it during any slot.

We have presented the first version of the formal specification of FlexRay in FOCUS and its verification in Isabelle/HOL were presented in [KS06a] and [Spi06] respectively. We have discussed the general introduction to the FlexRay formalization also in [KS06b] and [KSed]. To reduce the complexity of the system several aspects of FlexRay have been abstracted in this formalization:

(1) There is no clock synchronization or start-up phase since clocks are assumed to be synchronous. This corresponds very well with the *time-synchronous* notion of FOCUS.

(2) The model does not contain bus guardians that protect channels on the physical layer from interference caused by communication that is not aligned with FlexRay schedules.

(3) Only the static segment of the communication cycle has been included not the dynamic, as we are mainly interested in time-triggered systems.

(4) The time-basis for the system is one slot i.e. one slot FlexRay corresponds to one tick in in the formalization.

(5) The system contains only one FlexRay channel. Adding a second channel would mean simply doubling the FlexRay component with a different configuration and adding extra channels for the access to the *CNL_Buffer* component.

A formal verification of the clock synchronization algorithm and of the bus guardian of FlexRay is in progress at INRIA [Zha06].

The specification group *FR* consists of the following specifications, which describe the FlexRay components accordingly to the FlexRay standard [Fle04]:

    ✓ *FlexRay* (general requirements specification),
    ✓ *FlexRayArch* (system architecture),
    ✓ *FlexRayArchitecture* (guarantee part of the system architecture, for details
        see Section 4.2.5),
    ✓ *Cable*,
    ✓ *Controller*,
    ✓ *Scheduler*, and
    ✓ *BusInterface*.

We specify all these components as a joined Isabelle/HOL theory *FR.thy*(see
Section 3.4). We present the following Isabelle/HOL theories in this case study:

    ✓ *FR_types.thy* – the datatype definitions,
    ✓ *FR_inout.thy* – the specification of component interfaces and the corre-
        sponding correctness proofs,
    ✓ *FR.thy* – the Isabelle/HOL specifications of the system components and
        auxiliary functions and predicates,
    ✓ *FR_proof* – the proof of refinement relation between the requirements and
        the architecture specifications.

Here we present first of all meaning of the non-standard datatypes used in
the specification group, and the specification of the relations between sets of
(sheaves of) channels of the components, as well as proofs of their correctness
(see Section *FR_inout.thy*). Then we specify all the components from the specification group and auxiliary functions and predicates. After that we discuss
translation of the FOCUS component specifications into Isabelle/HOL, as well
as the proof of refinement relation between the requirements specification and
the architecture specification of FlexRay.

    The resulting Isabelle/HOL theories *FR.thy* and *FR_proof.thy* are presented in
Appendix B.

4.2.1. Datatypes

The type *Frame* that describes a FlexRay frame consists of a slot identifier of
type $\mathbb{N}$ and the payload. The type of payload is defined as a finite list of type
Message. The type *Config* represents the bus configuration and contains the
scheduling table *schedule* of a node and the length of the communication round
cycleLength. A scheduling table of a node consists of a number of slots in which
this node should be sending a frame with the corresponding identifier (identifier
that is equal to the slot).

> **type** *Message*   =   *msg* (*message_id* : $\mathbb{N}$, *ftcdata* : *Data*)
> **type** *Frame*   =   *frm* (*slot* : $\mathbb{N}$, *data* : *Data*)
> **type** *Config*   =   *conf* (*schedule* : $\mathbb{N}^*$, *cycleLength* : $\mathbb{N}$)

The Isabelle/HOL specifications of these types are equal modulo syntax to the
corresponding types in the FOCUS specification (see Section 2.3). We do not
specify the type *Data* here to have a polymorphic specification of FlexRay (this

type can be underspecified later to any datatype), therefore, in Isabelle/HOL it will be also defined as a polymorphic type $'a$.

The types $'a\ nFrame$, $nNat$ and $nConfig$ are used to represent sheaves of channels of types *Frame*, $\mathbb{N}$ and *Config* respectively (see Sections 2.3 and 2.13).

In the specification group will be used channels *recv* and *activations*, as well as sheaves of channels $(return_1, \ldots, return_n)$, $(c_1, \ldots, c_n)$, $(store_1, \ldots, store_n)$, $(get_1, \ldots, get_n)$, and $(send_1, \ldots, send_n)$. We also need to declare some constant, *sN*, for the number of specification replication and the corresponding number of channels in sheaves, as well as to define the list of sheaf upper bounds, *sheafNumbers*:

In the specification *FlexRayArchitecture* the stream *recv* is split into n streams (into a sheaf of channels). Therefore, we need to add this information to the Isabelle/HOL specification *FR_types.thy* – to add the definition of the *ch_split* predicate.

The Isabelle/HOL specification of these types and (component and channel) identifiers is presented below by the theory *FR_types.thy*.

theory *FR_types* = *Main* + *stream*:

record $'a\ Message$ =
  message_id :: *nat*
  ftcdata   :: $'a$

record $'a\ Frame$ =
  slot :: *nat*
  data :: $('a\ Message)\ list$

record *Config* =
  schedule   :: *nat list*
  cycleLength :: *nat*

types $'a\ nFrame = nat \Rightarrow ('a\ Frame)\ istream$
types $nNat = nat \Rightarrow nat\ istream$
types $nConfig = nat \Rightarrow Config$

consts *sN* :: *nat*

constdefs
  sheafNumbers :: *nat list*
  sheafNumbers $\equiv [sN]$

datatype *chanID* =
        ch_return
      | *ch_c*
      | *ch_store*
      | *ch_get*
      | *ch_send*
      | *ch_recv*
      | *ch_activation*

datatype *specID =*
       sFlexRay
      | *sFlexRayArch*
      | *sFlexRayArchitecture*
      | *sCable*
      | *sFlexRayController*
      | *sScheduler*
      | *sBusInterface*

datatype *csID = ch chanID* | *sheaf csID nat*

datatype *spID = spec specID* | *repl spID nat*

constdefs
  ch_split :: csID $\Rightarrow$ nat $\Rightarrow$ bool
  ch_split x l $\equiv$
  (x = ch ch_recv $\wedge$ l = sN)

end

4.2.2. Input/Output Relations between Channels

The Isabelle/HOL theory *FR_inout.thy* is based only on the Isabelle/HOL theory
inout_sheaf.thy (see Section 2.13.3), which is in the case of the specification group
FR based on the theory *FR_types.thy*. First of all we specify in this theory
the subcomponents relations for all components of the system by the function
subcomponents. Then we specify the list of input, output and local channels for
all components by the functions *ins*, *out* and *loc* respectively.

After the definition part we prove correctness of the interface relations showing that the predicates *correctInOutLoc*, *correctComposition*, *correctCompositionIn*,
correctCompositionOut, and *correctCompositionLoc* hold for the components of the
specification group by the standard proof schema (see Section 2.13.3).

theory *FR_inout = Main + inout_sheaf*:

primrec
  subcomponents sFlexRay = {}
  subcomponents sFlexRayArch = {spec sFlexRayArchitecture}
  subcomponents sFlexRayArchitecture
    = {repl (spec sFlexRayController) sN, spec sCable}
  subcomponents sCable   = {}
  subcomponents sFlexRayController
    = {spec sScheduler, spec sBusInterface}
  subcomponents sScheduler = {}
  subcomponents sBusInterface = {}

primrec
  subcomponentS (spec x) = subcomponents x
  subcomponentS (repl x y) = subcomponentS x

primrec
   ins sFlexRay = {*sheaf* (*ch ch_return*) *sN*}
   ins sFlexRayArch = {*sheaf* (*ch ch_return*) *sN*}
   ins sFlexRayArchitecture = {*sheaf* (*ch ch_return*) *sN*}
   ins sCable   = {*sheaf* (*ch ch_send*) *sN*}
   ins sFlexRayController = {*ch ch_return, ch ch_recv*}
   ins sScheduler   = {}
   ins sBusInterface = {*ch ch_activation, ch ch_recv, ch ch_return*}

primrec
   insS (*spec x*)   = *ins x*
   insS (*repl x i*) = (*cs2Sheaf i* ' (*insS x*))

primrec
   loc sFlexRay = {}
   loc sFlexRayArch = {}
   loc sFlexRayArchitecture = {*sheaf* (*ch ch_send*) *sN, ch ch_recv*}
   loc sCable = {}
   loc sFlexRayController = {*ch ch_activation*}
   loc sScheduler = {}
   loc sBusInterface = {}

primrec
   locS (*spec x*)   = *loc x*
   locS (*repl x i*) = (*cs2Sheaf i* ' (*locS x*))

primrec
   outS (*spec x*)   = *out x*
   outS (*repl x i*) = (*cs2Sheaf i* ' (*outS x*))

primrec
   out sFlexRay = {*sheaf* (*ch ch_store*) *sN, sheaf* (*ch ch_get*) *sN*}
   out sFlexRayArch = {*sheaf* (*ch ch_store*) *sN, sheaf* (*ch ch_get*) *sN*}
   out sFlexRayArchitecture = {*sheaf* (*ch ch_store*) *sN, sheaf* (*ch ch_get*) *sN*}
   out sCable   = {*ch ch_recv*}
   out sFlexRayController = {*ch ch_store, ch ch_get, ch ch_send*}
   out sScheduler   = {*ch ch_activation*}
   out sBusInterface = {*ch ch_store, ch ch_get, ch ch_send*}

Proofs for components

lemma *spec_FlexRay1*:
correctInOutLoc (*spec sFlexRay*)
  by (*simp add: correctInOutLoc_def*)

lemma *spec_FlexRay2*:
correctComposition (*spec sFlexRay*)
  by (*simp add: correctComposition_def*)

lemma *spec_FlexRayArch1* :
correctInOutLoc (*spec sFlexRayArch*)
  by (*simp add: correctInOutLoc_def*)

lemma *spec_FlexRayArch2* :
correctComposition (*spec sFlexRayArch*)
  by (*simp add: correctComposition_def*)

lemma *spec_FlexRayArch3* :
correctCompositionIn (*spec sFlexRayArch*)
  by (*simp add: correctCompositionIn_def*
        extSplit_def eqSplit_def)

lemma *spec_FlexRayArch4* :
correctCompositionOut (*spec sFlexRayArch*)
  by (*simp add: correctCompositionOut_def*
        extSplit_def eqSplit_def)

lemma *spec_FlexRayArch5* :
correctCompositionLoc (*spec sFlexRayArch*)
  by (*simp add: correctCompositionLoc_def eqSplit_def*
        split2loc_def)

lemma *spec_FlexRayArchitecture1* :
correctInOutLoc (*spec sFlexRayArchitecture*)
  by (*simp add: correctInOutLoc_def*)

lemma *spec_FlexRayArchitecture2* :
correctComposition (*spec sFlexRayArchitecture*)
  by (*simp add: correctComposition_def*)

lemma *spec_FlexRayArchitecture3* :
correctCompositionIn (*spec sFlexRayArchitecture*)
  by (*simp add: correctCompositionIn_def*
        extSplit_def ch_split_def
        sheafNumbers_def eqSplit_def
        cs2Sheaf_def split2sheaf_def makeSheafs_def, auto)

lemma *spec_FlexRayArchitecture4* :
correctCompositionOut (*spec sFlexRayArchitecture*)
  by (*simp add: correctCompositionOut_def*
        extSplit_def ch_split_def eqSplit_def
        cs2Sheaf_def split2sheaf_def makeSheafs_def, auto)

lemma *spec_FlexRayArchitecture5* :
correctCompositionLoc (*spec sFlexRayArchitecture*)
  by (*simp add: correctCompositionLoc_def*
        eqSplit_def ch_split_def split2loc_def
        cs2Sheaf_def, auto)

lemma *spec_Cable1*:
correctInOutLoc (spec sCable)
  by *(simp add: correctInOutLoc_def)*

lemma *spec_Cable2*:
correctComposition (spec sCable)
  by *(simp add: correctComposition_def)***lemma** *spec_FlexRayController1*:
correctInOutLoc (spec sFlexRayController)
  by *(simp add: correctInOutLoc_def)*

lemma *spec_FlexRayController2*:
correctComposition (spec sFlexRayController)
  by *(simp add: correctComposition_def)*

lemma *spec_FlexRayController3*:
correctCompositionIn (spec sFlexRayController)
  by *(simp add: correctCompositionIn_def*
        extSplit_def ch_split_def eqSplit_def split2sheaf_def)

lemma *spec_FlexRayController4*:
correctCompositionOut (spec sFlexRayController)
  by *(simp add: correctCompositionOut_def*
        extSplit_def ch_split_def eqSplit_def split2sheaf_def, auto)

lemma *spec_FlexRayController5*:
correctCompositionLoc (spec sFlexRayController)
  by *(simp add: correctCompositionLoc_def*
        *ch_split_def eqSplit_def split2loc_def, auto)***lemma** *spec_Scheduler1*:
correctInOutLoc (spec sScheduler)
  by *(simp add: correctInOutLoc_def)*

lemma *spec_Scheduler2*:
correctComposition (spec sScheduler)
  by *(simp add: correctComposition_def)***lemma** *spec_BusInterface1*:
correctInOutLoc (spec sBusInterface)
  by *(simp add: correctInOutLoc_def)*

lemma *spec_BusInterface2*:
correctComposition (spec sBusInterface)
  by *(simp add: correctComposition_def)*
end

⚠ **Remark:** To prove the lemma *spec_FlexRayArch5* we have used only the definitions of the predicates *correctCompositionLoc*, *eqSplit* and *split2loc*, because the specification *FlexRayArch* does not have any local channel, even though this specification is a composite one.

4.2.3. Auxiliary Predicates

In the specification of the group FR we will need three auxiliary predicates: *DisjointSchedules*, *IdenticCycleLength*, and *FrameTransmission*.

The predicate *DisjointSchedules* is true for sheaf of channels of type *Config*, if all bus configurations have disjoint scheduling tables:

```
┌─ DisjointSchedules ────────────────────────────────
│  c_1, ..., c_n ∈ Config
│  ─────────────────────────
│
│  ∀ i, j ∈ [1..n], j ≠ i :
│
│     ∀ x ∈ rng.schedule(c_i), y ∈ rng.schedule(c_j) :
│
│        x ≠ y
└────────────────────────────────
```

To represent this predicate in Isabelle/HOL we need to translate into Isabelle the expression

$$\forall x \in \mathsf{rng}.schedule(c_i), y \in \mathsf{rng}.schedule(c_j)$$

For this representation we define a new predicate *disjoint* in the Isabelle/HOL theory *ListExtras.thy* (see Appendix A.5). This predicate holds for two lists l_1 and l_2 of the same type, if the sets of elements of the lists l_1 and l_2 ($\mathsf{rng}.l_1$ and $\mathsf{rng}.l_1$) are pairwise disjoint.

According to the translation schema and the definition of the predicate *disjoint*, we get the following equalities (*mem* is here the predefined Isabelle/HOL operator "member of the list"):

$$[\![\mathsf{rng}.x]\!]_{Isab} \equiv finU\_range\ [\![x]\!]_{Isab} \equiv set\ [\![x]\!]_{Isab}$$

$$[\![\forall i \in \mathsf{rng}.l_1, j \in \mathsf{rng}.l_2 : i \neq j]\!]_{Isab}$$
$$\equiv \forall\ i\ mem\ (set\ [\![l_1]\!]_{Isab}),\ j\ mem\ (set\ [\![l_2]\!]_{Isab}).\ \ i \neq j$$
$$\equiv (set\ [\![l_1]\!]_{Isab}) \cap (set\ [\![l_2]\!]_{Isab}) = \{\}$$
$$\equiv disjoint\ [\![l_1]\!]_{Isab}\ [\![l_2]\!]_{Isab}$$

Therefore, we can write the following equality:

$$[\![\forall x \in \mathsf{rng}.schedule(c_i), y \in \mathsf{rng}.schedule(c_j) : x \neq y]\!]_{Isab}$$
$$\equiv disjiont\ [\![schedule(c_i)]\!]_{Isab}\ [\![schedule(c_j)]\!]_{Isab}$$
$$\equiv disjiont\ (schedule\ nC\ i)\ (schedule\ nC\ j)$$

We get the representation of the FOCUS predicate *DisjointSchedules* in Isabelle/HOL:

```
constdefs
   DisjointSchedules :: nat ⇒ nConfig ⇒ bool
   DisjointSchedules n nC
   ≡
   ∀ i j. i < n ∧ j < n ∧ i ≠ j ⟶
   disjoint (schedule (nC i)) (schedule (nC j))
```

The predicate *IdenticCycleLength* is true for sheaf of channels of type *Config*, if all bus configurations have the equal length of the communication round:

IdenticCycleLength
$c_1, ..., c_n \in Config$

$\forall\, i, j \in [1..n] :$

$\quad cycleLength(c_i) = cycleLength(c_j)$

The corresponding representation in Isabelle/HOL:

constdefs
IdenticCycleLength :: nat $\Rightarrow$ nConfig $\Rightarrow$ bool
IdenticCycleLength n nC
$\equiv$
$\forall\ i\ j.\ i < n \wedge j < n \longrightarrow$
cycleLength (nC i) = cycleLength (nC j)

The predicate *FrameTransmission* defines the correct message transmission: if the time t is equal modulo the length of the cycle (FlexRay communication round) to the element of the scheduler table of the node k, then this and only this node can send a data atn the tth time interval.

FrameTransmission
$store_1, ..., store_n, return_1, ..., return_n \in Frame^{\underline{\omega}}$
$get_1, ..., get_n \in \mathbb{N}^{\underline{\omega}}$
$c_1, ..., c_n \in Config$

$\forall\, t \in \mathbb{N}, k \in [1..n] :$
$\quad \text{let } s = \mathsf{mod}(t, cycleLength(c_k)) \text{ in}$
$\qquad s \in schedule(c_k) \rightarrow$

$\qquad\quad \mathsf{ti}(get_k, t) = \langle s \rangle \wedge$

$\qquad\quad \forall j \in [1..n], j \neq k : \mathsf{ti}(store_j, t) = \mathsf{ti}(return_k, t)$

The corresponding representation in Isabelle/HOL:

constdefs
FrameTransmission ::
$\quad nat \Rightarrow 'a\ nFrame \Rightarrow 'a\ nFrame \Rightarrow nNat \Rightarrow nConfig \Rightarrow bool$
FrameTransmission n nStore nReturn nGet nC
$\quad \equiv$
$\quad \forall\ (t{::}nat)\ (k{::}nat).\ k < n \longrightarrow$
$\qquad (\ \text{let } s = [\![\mathsf{mod}(t, cycleLength(c_k))]\!]_{Isab} \quad in$
$\qquad [\![s \in schedule(c_k)]\!]_{Isab} \longrightarrow (nGet\ k)\ t = [s]$
$\qquad\quad \wedge\ (\forall\ j.\ j < k \wedge j \neq n \longrightarrow ((nStore\ j)\ t) = ((nReturn\ k)\ t))\ )\,)$

where

$$[\![\mathsf{mod}(t,\,cycleLength(c_k))]\!]_{Isab} = t \bmod (cycleLength\ (nC\ k))$$
$$[\![s \in schedule(c_k)]\!]_{Isab} \equiv (s\ mem\ (schedule\ (nC\ k))$$

Thus, the whole Isabelle/HOL definition of the predicate *FrameTransmission* is

constdefs
  FrameTransmission ::
    nat $\Rightarrow$ *'a nFrame* $\Rightarrow$ *'a nFrame* $\Rightarrow$ *nNat* $\Rightarrow$ *nConfig* $\Rightarrow$ *bool*
  FrameTransmission n nStore nReturn nGet nC
  $\equiv$
  $\forall$ *(t::nat) (k::nat).* $k < n \longrightarrow$
  (*let* $s = t \bmod (cycleLength\ (nC\ k))$
   in
   ($s\ mem\ (schedule\ (nC\ k))$
   $\longrightarrow$
   $(nGet\ k\ t) = [s] \wedge$
   $(\forall\ j.\ j < n \wedge j \neq k \longrightarrow ((nStore\ j)\ t) = ((nReturn\ k)\ t))$))

The FOCUS predicate *Broadcast* describes properties of FlexRay broadcast. The schematical translation of this predicate into Isabelle/HOL predicate *Broadcast* is trivial.

_____Broadcast_____
$send_1, ..., send_n, recv \in Frame \stackrel{\omega}{=}$

$\forall t \in \mathbb{N}:$

  if $\exists k \in [1...n] : \mathsf{ti}(send_k, t) \neq \langle \rangle$

    then $\mathsf{ti}(recv, t) = \mathsf{ti}(send_k, t)$

    else $\mathsf{ti}(recv, t) = \langle \rangle$

  fi

constdefs
  Broadcast ::
    nat $\Rightarrow$ *'a nFrame* $\Rightarrow$ *'a Frame istream* $\Rightarrow$ *bool*
  Broadcast n nSend recv
  $\equiv$
  $\forall$ *(t::nat).*
  (*if* $\exists\ k.\ k < n \wedge ((nSend\ k)\ t) \neq []$
   then $(recv\ t) = ((nSend\ (SOME\ k.\ k < n \wedge ((nSend\ k)\ t) \neq []))\ t)$
   else $(recv\ t) = []$)

The predicates *Send* and *Receive* define the FOCUS relations on the streams to represent respectively data send and data receive by FlexRay controller. Their schematical translation into Isabelle/HOL predicates is also trivial.

Send

$return, send \in Frame^{\underline{\omega}};\ get, activation \in \mathbb{N}^{\underline{\omega}}$

$\forall\, t \in \mathbb{N}:$

    if $\text{ti}(activation, t) = \langle\rangle$

        then $\text{ti}(get, t) = \langle\rangle \wedge \text{ti}(send, t) = \langle\rangle$

        else $\text{ti}(get, t) = \text{ti}(activation, t) \wedge \text{ti}(send, t) = \text{ti}(return, t)$

    fi

---

constdefs

Send ::

'a Frame istream $\Rightarrow$ 'a Frame istream $\Rightarrow$ nat istream $\Rightarrow$ nat istream $\Rightarrow$ bool

Send return send get activation

$\equiv$

$\forall$ (t::nat).

(if (activation t) = []

then (get t) = [] $\wedge$ (send t) = []

else (get t) = (activation t) $\wedge$ (send t) = (return t))

Receive

$recv, store \in Frame^{\underline{\omega}};\ activation \in \mathbb{N}^{\underline{\omega}}$

$\forall\, t \in \mathbb{N}:$

    if $\text{ti}(activation, t) = \langle\rangle$

        then $\text{ti}(store, t) = \text{ti}(recv, t)$

        else $\text{ti}(store, t) = \langle\rangle$

    fi

---

constdefs

Receive ::

Frame istream $\Rightarrow$ Frame istream $\Rightarrow$ nat istream $\Rightarrow$ bool

Receive recv store activation

$\equiv$

$\forall$ (t::nat).

(if (activation t) = []

then (store t) = (recv t)

else (store t) = [])

4.2.4. Requirement Specification

The FOCUS specification *FlexRay* describes the interface of the FlexRay communication protocol and represents requirements on the protocol: If the scheduling tables are correct in terms of the predicates *DisjointSchedules* and *IdenticCycleLength* presented above, and also the FlexRay component receives in every time interval at most one message from each node (via channels $return_i$, $1 \leq i \leq n$), then

- ✓ the frame transmission by FlexRay must be correct in terms of the predicate *FrameTransmission* presented above,

- ✓ FlexRay component sends in every time interval at most one message to each node via channels get_i and $store_i$, $1 \leq i \leq n$).

FlexRay (constant $c_1, ..., c_n \in$ Config)	timed

in	$return_1, \ldots, return_n$: *Frame*
out	$store_1, \ldots, store_n$: *Frame*; $get_1, \ldots, get_n$: $\mathbb{N}$

asm	$\forall\, i \in [1..n]$: $\mathsf{msg}_1(return_i)$
	$DisjointSchedules(c_1, ..., c_n)$
	$IdenticCycleLength(c_1, ..., c_n)$
gar	$FrameTransmission(return_1, \ldots, return_n, store_1, \ldots, store_n, get_1, \ldots, get_n,$
	$c_1, \ldots, c_n)$
	$\forall\, i \in [1..n]$: $\mathsf{msg}_1(get_i) \wedge \mathsf{msg}_1(store_i)$

We convert the FOCUS specification *FlexRay* into Isabelle/HOL predicate *FlexRay* schematically.

constdefs

   FlexRay ::

   $nat \Rightarrow\ 'a\ nFrame \Rightarrow nConfig \Rightarrow\ 'a\ nFrame \Rightarrow nNat \Rightarrow bool$

   FlexRay n nReturn nC nStore nGet

   $\equiv$

   $(CorrectSheaf\ n)\ \wedge$

   $((\forall\ (i::nat).\ i < n \longrightarrow (msg\ 1\ (nReturn\ i)))\ \wedge$

   $(DisjointSchedules\ n\ nC) \wedge (IdenticCycleLength\ n\ nC)$

   $\longrightarrow$

   $(FrameTransmission\ n\ nStore\ nReturn\ nGet\ nC)\ \wedge$

   $(\forall\ (i::nat).\ i < n \longrightarrow (msg\ 1\ (nGet\ i)) \wedge (msg\ 1\ (nStore\ i)))\ )$

4.2.5. Architecture Specification

The architecture of the FlexRay communication protocol, which is a refinement of the specification *FlexRay* (this will be shown in Section 4.2.8), is specified as the FOCUS specification *FlexRayArch* that is an A/G one. The assumption part of the specification *FlexRayArch* is equal to the assumption part of the specification *FlexRay*. The guarantee part is represented by the specification *FlexRayArchitecture* (see below).

```
══ FlexRayArch (constant c₁, ..., cₙ ∈ Config)══════════════ timed ══
  in     return₁, ..., returnₙ : Frame
  out    store₁, ..., storeₙ : Frame;  get₁, ..., getₙ : ℕ

  asm    ∀ i ∈ [1..n] : msg₁(returnᵢ)
         DisjointSchedules(c₁, ..., cₙ)
         IdenticCycleLength(c₁, ..., cₙ)
  gar    (store₁, ..., storeₙ, get₁, ..., getₙ) :=
               FlexRayArchitecture(c₁, ..., cₙ)(return₁, dots, returnₙ)
```

The schematic translation of the FOCUS predicate *FlexRayArch* into Isabelle/HOL predicate *FlexRayArch* is straightforward:

```
constdefs
FlexRayArch :: nat ⇒ 'a nFrame ⇒ nConfig ⇒ 'a nFrame ⇒ nNat ⇒ bool
FlexRayArch n nReturn nC nStore nGet
≡
(CorrectSheaf n) ∧
( (∀ (i::nat). i < n ⟶ (msg 1 (nReturn i))) ∧
    (DisjointSchedules n nC) ∧ (IdenticCycleLength n nC) )
⟶
FlexRayArchitecture n nReturn nC nStore nGet )
```

The specification *FlexRayArchitecture* (see below) represents architecture of the FlexRay communication protocol without taking into account the refinement relation with the specification of the requirements. *FlexRayArchitecture* is a composite one and consists of the component *Cable* and n components *FlexRay_Controller* (for n nodes).

The FOCUS representation of the specification *FlexRayArchitecture* as plain text (constraint style):

```
══ FlexRayArchitecture (constant c₁, ..., cₙ ∈ Config)═════════ timed ══
  in     return₁, ..., returnₙ : Frame
  out    store₁, ..., storeₙ : Frame;  get₁, ..., getₙ : ℕ
  loc    send₁, ..., sendₙ : Frame;  recv : Frame

  (recv) := Cable(send₁, ..., sendₙ)
  ∀ i ∈ [1..n] : (storeᵢ, sendᵢ, getᵢ) := FlexRayController(cᵢ)(returnᵢ, recv)
```

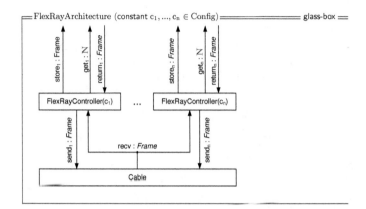

We convert the FOCUS specification *FlexRayArchitecture* into Isabelle/HOL predicate *FlexRayArchitecture* schematically:

constdefs
FlexRayArchitecture ::
    $nat \Rightarrow$ 'a $nFrame \Rightarrow nConfig \Rightarrow$ 'a $nFrame \Rightarrow nNat \Rightarrow bool$
FlexRayArchitecture n nReturn nC nStore nGet
$\equiv$
(CorrectSheaf n) $\wedge$
$(\exists\ nSend\ recv.$
    (Cable n nSend recv) $\wedge$
    $(\forall\ i.\ i < n \longrightarrow$
    FlexRayController (nReturn i) recv (nC i) (nStore i) (nSend i) (nGet i)) $)$

4.2.6. Cable Component

The FOCUS specification *Cable* describes properties of FlexRay broadcast and is an A/G specification. The component *Cable* simulate the broadcast properties of the physical network cable – every received FlexRay frame is resent to all connected nodes. Thus, if one *FlexRayController* send some frame, this frame will be resent to all nodes (to all *FlexRayControllers* of the system).

The assumption is that all input streams of the component *Cable* are disjoint – this holds by the properties of the *FlexRayController* components and the overall system assumption that the scheduling tables of all nodes are disjoint. This assumption is expressed using the FOCUS operator disj^{inf} for sheaves of channels (see Section 2.13.4). The guarantee is specified by the predicate *Broadcast* defined in Section 4.2.3. The schematic translation of the FOCUS predicate *Cable* into Isabelle/HOL predicate *Cable* is straightforward.

```
┌══ Cable ═══════════════════════════════════════════ timed ══
│  in     send₁, ..., sendₙ : Frame
│
│  out    recv : Frame
│ ──────────────────────────────────────────────────────────
│  asm    disjⁱⁿᶠ(send₁, ..., sendₙ)
│ ─ ─ ─ ─ ─ ─ ─ ─ ─ ─ ─ ─ ─ ─ ─ ─ ─ ─ ─ ─ ─ ─ ─ ─ ─ ─ ─ ─
│  gar    Broadcast(send₁, ..., sendₙ, recv)
```

constdefs

Cable :: nat ⇒ 'a nFrame ⇒ 'a Frame istream ⇒ bool
Cable n nSend recv
≡
(CorrectSheaf n) ∧ (inf_disj n nSend ⟶ Broadcast n nSend recv)

4.2.7. Controller Component

The FOCUS specification *FlexRayController* represent the controller component for a single node of the system. This specification is a composite one and consists of the components *Scheduler* and *BusInterface*. The *Scheduler* signals the *BusInterface*, that is responsible for the interaction with other nodes of the system, on which time which FlexRay frames must be send from the node.

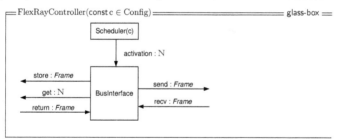

The FOCUS representation of the specification *FlexRayController* as plain text (constraint style):

```
┌══ FlexRayController(const c ∈ Config) ═══════════════ timed ══
│  in     return, recv : Frame
│
│  out    store, send : Frame; get : ℕ
│
│  loc    activation : ℕ
│ ──────────────────────────────────────────────────────────
│  (activation) := Scheduler(c)
│  (store, send, get) := BusInterface(activation, return, recv)
```

The schematic translation of the FOCUS predicate *FlexRayController* into Isabelle/HOL predicate *FlexRayController*:

> **constdefs**
> *FlexRayController ::*
>   *'a Frame istream ⇒ 'a Frame istream ⇒ Config ⇒*
>   *'a Frame istream ⇒ 'a Frame istream ⇒ nat istream ⇒ bool*
> *FlexRayController return recv c store send get*
> ≡
> *(∃ activation. (Scheduler c activation) ∧*
>   *(BusInterface activation return recv store send get))*

The *Scheduler* describes the communication scheduler. It sends at every time t interval, which is equal modulo the length of the communication cycle to some FlexRay frame identifier (that corresponds to the number of the slot in the communication round) from the scheduler table, this frame identifier.

```
══ Scheduler(const c ∈ Config) ═══════════════════════════ timed ══
 out   activation : ℕ
───────────────────────────────────────────────────────────────
 univ  s ∈ ℕ
 - - - - - - - - - - - - - - - - - - - - - - - - - - - - - - - -
 ∀ t ∈ ℕ :
    s = mod(t, cycleLength(c))
    if  s ∈ schedule(c)
       then  ti(activation, t) = ⟨s⟩
       else  ti(activation, t) = ⟨⟩
    fi
```

We convert the FOCUS specification *Scheduler* into Isabelle/HOL predicate *Scheduler* schematically:

> **constdefs**
> *Scheduler :: Config ⇒ nat ⇒ bool*
> *Scheduler c activation*
> ≡
> *∀ (t::nat).*
>   *(let s = t mod (cycleLength c)*
>   *in*
>   *(if s mem (schedule c)*
>   *then activation t = [s]*
>   *else activation t = []))*

The component *BusInterface* is responsible for the real send and receive of frames. We specify this component using by two predicates, *Send* and *Receive*. The schematic translation of the FOCUS specification *BusInterface* into Isabelle/HOL predicate is straightforward.

╒══ BusInterface ══════════════════════════════════ timed ══

in   *activation* : ℕ; *return, recv* : *Frame*

out   *send, store* : *Frame*; *get* : ℕ

───

Receive(*recv, store, activation*)

Send(*return, send, get, activation*)
---

constdefs
  BusInterface ::
      nat istream ⇒ *'a Frame istream* ⇒ *'a Frame istream* ⇒
      'a Frame istream ⇒ *'a Frame istream* ⇒ *nat istream* ⇒ *bool*
  BusInterface activation return recv store send get
  ≡
  (*Receive recv store activation*) ∧ (*Send return send get activation*)

4.2.8. Verification of the FlexRay System wrt. its Requirements

To show that the specified system fulfill the requirements we need to show that the specification *FlexRayArch* is a refinement of the specification *FlexRay*. It follows from the definition of behavioral refinement that in order to verify that

$$FlexRay \rightsquigarrow FlexRayArch \tag{4.3}$$

it is enough to prove that

$$[\![FlexRayArch]\!] \Rightarrow [\![FlexRay]\!] \tag{4.4}$$

Therefore, we have to define and to prove a *lemma*, that says the specification *FlexRayArch* is a refinement of the specification *FlexRay* (let us call this lemma *main_fr_refinement*):

lemma *main_fr_refinement*:
FlexRayArch n nReturn nC nStore nGet ⟹ *FlexRay n nReturn nC nStore nGet*

To prove this lemma we used first of all the definition of the predicates *FlexRayArch*, *FlexRay*, *FlexRayArchitecture*, and *CorrectSheaf*. After that we split the goal into three subgoals using the Isabelle/HOL automatic proof strategy:

1. The predicate *FrameTransmission* holds, if

✓ every stream of the sheaf *nReturn* has at every time interval at most one message;

✓ the number *n* of channels in the sheaf is greater then zero;

✓ the predicate *Cable* holds for the corresponding streams;

✓ the predicate *FlexRayController* holds the corresponding streams on every node *i* of the *n* nodes;

✓ the predicates *DisjointSchedules* and *IdenticCycleLength* hold for a sheaf of parameters *nC*.

[∀i < n. msg (Suc 0) (nReturn i); 0 < n; DisjointSchedules n nC;
IdenticCycleLength n nC; Cable n nSend recv;
∀i<n. FlexRayController (nReturn i) recv (nC i) (nStore i) (nSend i) (nGet i)]
⟶ FrameTransmission n nStore nReturn nGet nC

2. The *i*th stream of the sheaf *nGet* has at every time interval at most one message, if

✓ the index *i* is less then the number of channels in the sheaf;

✓ every stream of the sheaf *nReturn* has at every time interval at most one message;

✓ the number *n* of channels in the sheaf is greater then zero;

✓ the predicate *Cable* holds for the corresponding streams;

✓ the predicate *FlexRayController* holds the corresponding streams on every node *i* of the *n* nodes;

✓ the predicates *DisjointSchedules* and *IdenticCycleLength* hold for a sheaf of parameters *nC*.

[∀i < n. msg (Suc 0) (nReturn i); DisjointSchedules n nC;
IdenticCycleLength n nC; Cable n nSend recv; i < n;
∀i<n. FlexRayController (nReturn i) recv (nC i) (nStore i) (nSend i) (nGet i)]
⟶ msg (Suc 0) (nGet i)

3. The *i*th stream of the sheaf *nStore* has at every time interval at most one message, if

✓ the index *i* is less then the number of channels in the sheaf;

✓ every stream of the sheaf *nReturn* has at every time interval at most one message;

✓ the number *n* of channels in the sheaf is greater then zero;

✓ the predicate *Cable* holds for the corresponding streams;

✓ the predicate *FlexRayController* holds the corresponding streams on every node *i* of the *n* nodes;

✓ the predicates *DisjointSchedules* and *IdenticCycleLength* hold for a sheaf of parameters *nC*.

$\llbracket \forall i < n.\ msg\ (Suc\ 0)\ (nReturn\ i);\ DisjointSchedules\ n\ nC;$
$IdenticCycleLength\ n\ nC;\ Cable\ n\ nSend\ recv;\ i < n;$
$\forall i < n.\ FlexRayController\ (nReturn\ i)\ recv\ (nC\ i)\ (nStore\ i)\ (nSend\ i)\ (nGet\ i)\rrbracket$
$\longrightarrow msg\ (Suc\ 0)\ (nStore\ i)$

We define a lemma *fr_refinement_FrameTransmission* (see below) which is equal to the first subgoal and prove it externally. Thus, the first subgoal can be solved by lemma *fr_refinement_FrameTransmission*.

The last two subgoals have the same assumption part and differ only in their goals, which also have a similar structure. We define a lemma *fr_refinement_msg* (see below) which assumption part is equal to the assumption parts of these subgoals and the goal is the conjunction of their goals. Thus, the second and the third subgoal can be solved by lemma *fr_refinement_msg*.

When the proof of a new lemma succeeds, we can test, which of its assumptions are real necessary and which of them can be removed. E.g., in the case of lemma *fr_refinement_FrameTransmission* the assumption

$$\forall i < n.\ msg\ (Suc\ 0)\ (nReturn\ i)$$

can be removed. Thus, the lemma *fr_refinement_FrameTransmission* has the following form:

The predicate *FrameTransmission* holds, if

✓ the number n of channels in the sheaf is greater then zero;

✓ the predicate *Cable* holds for the corresponding streams;

✓ the predicate *FlexRayController* holds the corresponding streams on every node i of the n nodes;

✓ the predicates *DisjointSchedules* and *IdenticCycleLength* hold for a sheaf of parameters nC.

lemma *fr_refinement_FrameTransmission*:
$\llbracket$ *Cable* n *nSend recv*; $0 < n$;
$\forall i < n.$ *FlexRayController* (*nReturn* i) *recv* (*nC* i) (*nStore* i) (*nSend* i) (*nGet* i);
DisjointSchedules n *nC*; *IdenticCycleLength* n *nC* $\rrbracket$
$\Longrightarrow$ *FrameTransmission* n *nStore nReturn nGet nC*

To prove the lemma we use first the definition of the predicate *FrameTransmission* and the *let*-expression, the Isabelle/HOL automatic proof strategies, and the lemma *fr_nStore_nReturn* (see below).

After that we get a goal, where the argumentation is over the kth node. Therefore, we instantiate the $\forall$ quantifier (over the index of a node) with k. Then we apply the definition of the predicates *FlexRayController*, *BusInterface*, *Send* and *Scheduler*, as well as the Isabelle/HOL automatic proof strategies.

To solve the resulting goal we instantiate all the $\forall$ quantifiers with t (number of the time interval) and apply the definition of the *let*-expression.

The lemma *fr_nStore_nReturn* says, that at the time interval t the list of messages of the output stream *store* of any jth node ($j \neq k$) is equal to the list of messages of the input stream *return* of the node k, if

✓ the number n is greater then zero and the number k is less then n;

✓ the predicate *Cable* holds for the corresponding streams;

✓ the time t modulo the length of the communication cycle is an element of the communication scheduling table of the kth node;

✓ for a sheaf of parameters nC hold the predicates *DisjointSchedules* and *IdenticCycleLength*;

✓ the predicate *FlexRayController* holds for the corresponding streams on every node i of the n nodes.

lemma *fr_nStore_nReturn*:
⟦*Cable n nSend recv*;
 $\forall i < n$. *FlexRayController* (*nReturn i*) *recv* (*nC i*) (*nStore i*) (*nSend i*) (*nGet i*);
 $0 < n$; $k < n$; *DisjointSchedules n nC*; *IdenticCycleLength n nC*;
 t *mod cycleLength* (*nC k*) *mem schedule* (*nC k*)⟧
 $\implies \forall j. \ j < n \wedge j \neq k \longrightarrow nStore \ j \ t = nReturn \ k \ t$

To prove the lemma we first of all use the Isabelle/HOL automatic proof strategy and add new assumption according lemma *fr_nC_Send* (see below), which says that then at the time interval t the list of messages of the output stream *send* of any jth node ($j \neq k$) must be empty.

After that we apply the definitions of the predicates *Cable* and *CorrectSheaf*, as well as add new assumption according the lemma *disjointFrame_lemma* (see below), which says that all streams in the sheaf *nSend* are disjoint.

To argue at next steps about two nodes, j and k, of the system we duplicate the assumption about the FlexRay controller and instantiate the $\forall$ quantifiers in the duplicated assumption with j and k.

After that we apply the definition of the predicate *Broadcast* and instantiate the $\forall$ quantifiers in the definition of the *Broadcast* component with the number of the time interval t.

As the next proof steps we apply the definitions of the predicates *FlexRay-Controller* and *Scheduler* and use the Isabelle/HOL automatic proof strategies. Then we instantiate the $\forall$ quantifiers in the definition of the *Scheduler* component with the number of the time interval t, and apply the definition of the *let*-expression.

Now we can split the current goal on two cases: whether the time t modulo the length of the communication cycle is an element of the communication scheduling table of the jth node. The first goal can be solved by lemma *correct_DisjointSchedules1* (see below). To solve the second goal we apply the definitions of the predicates *BusInterface* and *Send*, instantiate the $\forall$ quantifiers in the definition of the *Send* component with the number of the time interval t, and split split the if_then_else-expressions.

To solve the resulting goals we apply the lemma *inf_disj_index* (property of disjoint channels in a sheaf, see Section A.1), the definitions of the predicates *inf_disj* and *Receive*, as well as use the Isabelle/HOL automatic proof strategies.

The lemma *fr_nC_Send* says, that at any time interval t the list of messages of the output stream *send* of any jth node ($j \neq k$) must be empty, if

✓ the predicate *FlexRayController* holds for the corresponding streams on every node i of the n nodes;

✓ the number n is greater then zero and the number k is less then n;

✓ for a sheaf of parameters nC hold the predicates *DisjointSchedules* and *IdenticCycleLength*;

✓ the time t modulo the length of the communication cycle is an element of the communication scheduling table of the kth node.

lemma *fr_nC_Send*:
⟦ ∀ $i < n$. *FlexRayController* ($nReturn\ i$) $recv$ ($nC\ i$) ($nStore\ i$) ($nSend\ i$) ($nGet\ i$);
$0 < n$; $k < n$;
DisjointSchedules $n\ nC$; *IdenticCycleLength* $n\ nC$;
$t\ mod\ cycleLength$ ($nC\ k$) $mem\ schedule$ ($nC\ k$)⟧
⟹
∀ j. $j < n \wedge j \neq k \longrightarrow$ ($nSend\ j$) $t = []$

To prove this lemma we use first of all use the Isabelle/HOL automatic proof strategy and add new assumption according the lemma *correct_DisjointSchedules1* (see below), which says that the time t modulo the length of the communication cycle is not an element of the communication scheduling table of the jth node. To solve the resulting goal we instantiate the ∀ quantifier with j (index of the treated node) and apply the lemma *fr_Send* (see below).

The lemma *fr_Send* says, that at any time interval t the list of messages of the output stream *send* of any ith node must be empty, if

✓ the predicate *FlexRayController* holds for the corresponding streams on a node i;

✓ the time t modulo the length of the communication cycle is an element of the communication scheduling table of the ith node.

lemma *fr_Send*:
⟦ *FlexRayController* ($nReturn\ i$) $recv$ ($nC\ i$) ($nStore\ i$) ($nSend\ i$) ($nGet\ i$);
¬ ($t\ mod\ cycleLength$ ($nC\ i$) $mem\ schedule$ ($nC\ i$))⟧
⟹ ($nSend\ i$) $t = []$

To prove this lemma we apply first of all the definitions of the predicates *FlexRayController* and *Scheduler*, and use the Isabelle/HOL automatic proof strategy.

As the next steps we instantiate the ∀ quantifier with *t*, and apply the definition of the *let*-expression and of the predicates *BusIntefaceof* and *Send*, as well as the Isabelle/HOL automatic proof strategy.

The lemma *correct_DisjointSchedules1* says, that at any time interval *t*, *t* modulo the length of the communication cycle cannot be an element of the communication scheduling table of the *j*th node, if

✓ the number *n* is greater then zero, and the numbers *k* and *j* are less then *n* (*k* ≠ *j*);

✓ *t*, the index of the time interval modulo the length of the communication cycle is an element of the communication scheduling table of the *k*th node;

✓ the predicates *DisjointSchedules* and *IdenticCycleLength* hold for a sheaf of parameters *nC*;

lemma *correct_DisjointSchedules1*:
⟦ *DisjointSchedules n nC*; *IdenticCycleLength n nC*;
(*t mod cycleLength* (*nC k*)) *mem schedule* (*nC k*); *k* < *n*; *j* < *n*; *k* ≠ *j* ⟧
⟹ ¬ (*t mod cycleLength* (*nC j*) *mem schedule* (*nC j*))

To prove this lemma we apply first of all the definition of the predicate *Disjoint-Schedules* and we instantiate the ∀ quantifiers with the indexes of the nodes *k* and *j*. After that we use the Isabelle/HOL automatic proof strategy, apply the definition of the predicate *IdenticCycleLength* and instantiate the ∀ quantifiers with the indexes of the nodes *k* and *j* again.

To solve the resulting goal we add new assumption according the lemma *mem_notdisjoint* (from the Isabelle/HOL theory *ListExtras.thy*, see Section A.5), which says that the communication scheduling tables of the nodes *k* and *j* are not disjoint, and use the Isabelle/HOL automatic proof strategy.

The lemma *disjointFrame_lemma* says, that all the streams of the sheaf *nSend* are disjont, if

✓ the number *n* is greater then zero;

✓ the predicates *DisjointSchedules* and *IdenticCycleLength* hold for the sheaf of parameters *nC*;

✓ the predicate *FlexRayController* holds for the corresponding streams on every node *i* of the *n* nodes.

lemma *disjointFrame_lemma*:
⟦ *DisjointSchedules n nC*; *0* < *n*; *IdenticCycleLength n nC*;
∀ *i* < *n*. *FlexRayController* (*nReturn i*) *rcv* (*nC i*) (*nStore i*) (*nSend i*) (*nGet i*) ⟧
⟹ *inf_disj n nSend*

To argue at next steps about two nodes of the system we duplicate the assumption about the FlexRay controller. Then we apply the definition of the predicate *inf_disj*, use the Isabelle/HOL automatic proof strategy and instantiate the $\forall$ quantifiers with the indexes of the nodes k and j.

After that we apply the definitions of the predicates *FlexRayController* and *DisjointSchedules*, instantiate the $\forall$ quantifiers with the indexes of the nodes k and j again, apply the definitions of the predicates *BusInterface* and *Send*.

As the next step we instantiate all the $\forall$ quantifiers with the index of the current time interval, t, apply the definitions of the *if_then_else* expression and the predicate *Scheduler*. Then we instantiate all the $\forall$ quantifiers with t, and apply the definitions of the *if_then_else* and *let* expressions. As the next step we apply the definition of the predicate *IdenticCycleLength*.

To solve the resulting goal we instantiate the $\forall$ quantifiers with indexes of nodes i and j, and show that the assumptions contain contradiction.

The lemma *fr_refinement_msg* says, that the streams get_i and $store_i$ must contain at every time interval at most one FlexRay frame, if

✓ the number n is greater then zero and the number i is less then n;

✓ the predicates *Cable* and *FlexRayController* hold for the corresponding streams;

✓ the streams $return_1, \ldots, return_n$ contain at every time interval at most one FlexRay frame;

✓ for a sheaf of parameters nC hold the predicates *DisjointSchedules* and *IdenticCycleLength*;

lemma *fr_refinement_msg*:
⟦ *Cable n nSend recv*; *DisjointSchedules n nC*; *IdenticCycleLength n nC*;
  $i < n$; $0 < n$; $\forall i{<}n.$ *msg* $(Suc\ 0)$ $(nReturn\ i)$;
  $\forall i{<}n.$ *FlexRayController* $(nReturn\ i)$ *recv* $(nC\ i)$ $(nStore\ i)$ $(nSend\ i)$ $(nGet\ i)$⟧
  $\Longrightarrow$ *msg* $(Suc\ 0)$ $(nGet\ i)$ $\wedge$ *msg* $(Suc\ 0)$ $(nStore\ i)$

To prove the lemma we prove first that the streams of the sheaf *nSend* are disjoint (an additional subgoal "*inf_disj n nSend*") using an auxiliary lemma *disjointFrame_lemma*. After that we split the goal by the conjunction rule in two subgoals und solve them using lemmas *fr_refinement_msg_nGet* and *fr_refinement-msg_nStore* (see below), as well as the Isabelle/HOL automatic proof strategy.

The lemma *fr_refinement_msg_nGet* says, that the stream get_i must contain at every time interval at most one FlexRay frame, if

✓ the number i is less then the number n;

✓ the predicate *FlexRayController* holds for the corresponding streams on every node i of the n nodes.

lemma *fr_refinement_msg_nGet*:
⟦ $\forall i{<}n.$ *FlexRayController* $(nReturn\ i)$ *recv* $(nC\ i)$ $(nStore\ i)$ $(nSend\ i)$ $(nGet\ i)$;
  $< n$⟧ $\Longrightarrow$ *msg* $(Suc\ 0)$ $(nGet\ i)$

To prove this lemma we apply first of all the definition of the predicate *FlexRay-Controller* and instantiate all the ∀ quantifier with i, use the Isabelle/HOL automatic proof strategy and apply the definitions of the predicates *BusInterface*, *msg*, *Send* and *Scheduler*.

To solve the resulting goal we instantiate all the ∀ quantifiers with the index of the current time interval, t, and apply the definitions of the *let* and *if_then_else* expressions.

The lemma *fr_refinement_msg_nStore* says, that the stream $store_i$ must contain at every time interval at most one FlexRay frame, if

✓ for a sheaf of parameters nC hold the predicates *DisjointSchedules* and *IdenticCycleLength*;

✓ the streams $return_1, \ldots, return_n$ contain at every time interval at most one FlexRay frame;

✓ the predicate *Cable* holds for the corresponding streams;

✓ the streams of the sheaf *nSend* are disjoint;

✓ the predicate *FlexRayController* holds for the corresponding streams on every node i of the n nodes;

lemma *fr_refinement_msg_nStore*:
⟦ *DisjointSchedules* n nC; *IdenticCycleLength* n nC;
inf_disj n $nSend$; $i < n$; $0 < n$;
∀ $i<n.$ *msg* $(Suc\ 0)$ $(nReturn\ i)$; *Cable* n $nSend$ $recv$;
∀ $i<n.$ *FlexRayController* $(nReturn\ i)$ $recv$ $(nC\ i)$ $(nStore\ i)$ $(nSend\ i)$ $(nGet\ i)$⟧
⟹ *msg* $(Suc\ 0)$ $(nStore\ i)$

To prove this lemma we apply first of all the definitions of the predicates *msg*, *Cable*, *CorrectSheaf* and *Broadcast*, instantiate the ∀ quantifier with t in the last assumption, and apply the the definition of the *if_then_else* expression. As result we get two cases:

✓ The case "kth stream of the sheaf *nSend* is nonempty at time interval t".

To solve this goal we add first new assumption according the lemma *inf_disj_index* (property of disjoint channels in a sheaf, see Section A.1), use the Isabelle/HOL automatic proof strategy and apply the definition of the predicate *inf_disj*.

After that we instantiate the ∀ quantifiers with t, the index of the current time interval, and k, the index of the stream which is currently nonempty, and duplicate the assumption about FlexRay controllers (to argue later about two nodes of the system, k and i). Then we instantiate the ∀ quantifiers with k and i in the assumptions about FlexRay controller and apply the definition of the corresponding predicate.

As the next step we instantiate the $\forall$ quantifiers with k, and add new assumption according the lemma *fr_refinement_msg_nSend* (see above), which says that the $k$the stream of the sheaf *nSend* contains at most one message in each time interval.

After applying the definitions of the predicates *BusInterface* and *Receive*, we instantiate the $\forall$ quantifiers with t in the assumptions about the stream *activation*, apply the the definition of the *if_then_else* expression of the predicate *msg*.

✓ The case "All streams of the sheaf *nSend* are empty at time interval t".

To solve this goal we apply the definitions of the predicates *FlexRay-Controller*, *BusInterface* and *Receive*, as well as instantiation rules, the definition of the *if_then_else* expression and the Isabelle/HOL automatic proof strategy.

The lemma *fr_refinement_msg_nSend* says, that the stream $send_i$ must contain at every time interval at most one FlexRay frame, if

✓ the streams $return_1, \ldots, return_n$ contain at every time interval at most one FlexRay frame;

✓ for a sheaf of parameters nC hold the predicates *DisjointSchedules* and *IdenticCycleLength*;

✓ the predicate *BusInterface* is true for the corresponding streams on the node i.

lemma *fr_refinement_msg_nSend*:
⟦ *msg* (*Suc* 0) (*nReturn i*);
   BusInterface activation (*nReturn i*) *recv* (*nStore i*) (*nSend i*) (*nGet i*)⟧
⟹ *msg* (*Suc* 0) (*nSend i*)

We prove this lemma using the definitions of the predicates *msg*, *BusInterface* and *Send*, as well as instantiation rules, the definition of the *if_then_else* expression, and the Isabelle/HOL automatic proof strategy.

The whole proofs of the system properties are represented in the Appendix B.4.

4.2.9. Results of the Case Study

In this case study we have shown how we can deal with sheaves of channels ans parameters, as well as with specification replications.

The FOCUS specifications of all components of the FlexRay system were translated schematically to Isabelle/HOL and the refinement relation between the requirement and the architecture specification of the system was proved.

The correctness of the input/output relations was also proved for all components of the system.

4.3. Automotive-Gateway

This section introduces the case study on telematics (electronic data transmission) gateway that was done for the Verisoft project (see [Ver] and [Autb]). If the gateway receives (from a ECall application of a vehicle) a signal about crash (more precise, the command to initiate the call to the Emergency Service Center), and after the establishing the connection it receives the command to send the crash data (these data were already received and stored in the internal buffer of the gateway), these data will be resent to the Emercency Service Center and the voice communication will be established, assuming that there is no connection fails.

The specification group *Gateway* consists of the following components components:

✓ *GatewaySystem* (gateway system architecture),

✓ *GatewaySystemReq* (gateway system requirements),

✓ *ServiceCenter* (Emergency Service Center),

✓ *Gateway* (gateway architecture),

✓ *GatewayReq* (gateway requirements),

✓ *Sample* (the main subcomponent of the gateway, which describes its logic),

✓ *Delay* (the subcomponent of the *Gateway* to model the communication delay), and

✓ *Loss* (the subcomponent of the *Gateway* to model the communication loss).

We specify all these components as a joined Isabelle/HOL theory (see Section 3.4). We present the following Isabelle/HOL theories in this case study:

✓ *Gateway_types.thy* – the datatype definitions,

✓ *Gateway_inout.thy* – the specification of component interfaces and the corresponding correctness proofs,

✓ *Gateway.thy* – the Isabelle/HOL specifications of the system components,

✓ *Gateway_proof* – the proofs of refinement relations between the requirements and the architecture specifications (for the components *Gateway* and *GatewaySystem*), as well as the equality proof of the compressed and simple translations of the tiTable (see Section 2.6).

Here we present first of all meaning of the non-standard datatypes used in the specification group and the specification of the relations between sets of channels of the components, as well as proofs of their correctness (see Section *Gateway_inout.thy*). Then we specify all the components from the specification group and auxiliary functions and predicates. After that we discuss translation of the FOCUS component specifications into Isabelle/HOL, as well as the proof of refinement relations between the requirements specifications and the architecture specifications.

The resulting Isabelle/HOL theories *Gateway.thy* and *Gateway_proof.thy* are presented in Appendix B.

4.3.1. Datatypes

The datatype *ECall_Info* represents a tuple, consisting of the data that the Emergency Service Center needs – here we specify these data to contain the vehicle coordinates and the collision speed, they can also extend by some other information. The datatype *GatewayStatus* represents the status (internal state) of the gateway.

$$
\begin{aligned}
\textbf{type } Coordinates \quad &= \quad \mathbb{N} \times \mathbb{N} \\
\textbf{type } CollisionSpeed \quad &= \quad \mathbb{N} \\
\textbf{type } ECall\_Info \quad &= \quad ecall(coord \in Coordinates, speed \in CollisionSpeed) \\
\textbf{type } GatewayStatus \quad &= \quad \{ \ init\_state, \ call, \ connection\_ok, \\
& \qquad sending\_data, \ voice\_com \ \}
\end{aligned}
$$

The Isabelle/HOL specifications of these types are equal modulo syntax to the corresponding types in the FOCUS specification (see Section 2.3).

To specify the automotive gateway we will use a number of datatypes consisting of one or two elements: $\{init, send\}$, $\{stop\_vc\}$, $\{vc\_com\}$ and $\{sc\_ack\}$. We name these types *reqType*, *stopType*, *vcType* and *aType* correspondingly, and represent them in Isabelle/HOL schematically (see Section 2.3). The Isabelle/HOL specification of these types as well as of the component and channel identifiers is presented below by the theory *Gateway_types.thy*.

theory *Gateway_types = Main + stream:*

types
    Coordinates = nat × nat
types
    CollisionSpeed = nat

record *ECall_Info =*
    coord :: Coordinates
    speed :: CollisionSpeed

datatype *GatewayStatus =*
    init_state
    | *call*
    | *connection_ok*
    | *sending_data*
    | *voice_com*

datatype *reqType* = *init* | *send*

datatype *stopType* = *stop_vc*

datatype *vcType* = *vc_com*

datatype *aType* = *sc_ack*

datatype *chanID* =
      ch_req
    | *ch_dt*
    | *ch_ack*
    | *ch_stop*
    | *ch_lose*
    | *ch_i*
    | *ch_a*
    | *ch_vc*
    | *ch_a1*
    | *ch_a2*
    | *ch_i1*
    | *ch_i2*

datatype *specID* =
      sGatewaySystem
    | *sGatewaySystemReq*
    | *sServiceCenter*
    | *sGatewayReq*
    | *sGateway*
    | *sLoss*
    | *sDelay*
    | *sSample*

end

4.3.2. Input/Output Relations between Channels

The Isabelle/HOL theory *Gateway_inout.thy* is based only on the Isabelle/HOL theory *inout.thy* (see Section 2.11.4), which is in the case of the specification group *Gateway* based on the theory *Gateway_types.thy*. First of all we specify in this theory the subcomponents relations for all components of the system by the function *subcomponents*. Then we specify the sets of input, output and local channels for all components by the functions *ins*, *out* and *loc* respectively. After that we prove that the predicates *correctInOutLoc* and *correctComposition* hold for all components, and also that the predicates *correctCompositionIn*, *correctCompositionOut* and *correctCompositionLoc* holds for composite components.

theory *Gateway_inout = Main + inout*:

primrec
  subcomponents sGatewaySystem =
        {sGateway, sServiceCenter}
  subcomponents sGatewaySystemReq = {}
  subcomponents sServiceCenter   = {}
  subcomponents sGatewayReq   = {}
  subcomponents sGateway = {sSample, sDelay, sLoss}
  subcomponents sLoss   = {}
  subcomponents sDelay = {}
  subcomponents sSample = {}

primrec
  ins sGatewaySystem =
        {ch_req, ch_dt, ch_stop, ch_lose}
  ins sGatewaySystemReq =
        {ch_req, ch_dt, ch_stop, ch_lose}
  ins sServiceCenter   = {ch_i}
  ins sGatewayReq   =
        {ch_req, ch_dt, ch_stop, ch_lose, ch_a}
  ins sGateway =
        {ch_req, ch_dt, ch_stop, ch_lose, ch_a}
  ins sLoss   = {ch_a, ch_i2, ch_lose}
  ins sDelay = {ch_a2, ch_i1}
  ins sSample =
        {ch_req, ch_dt, ch_stop, ch_lose, ch_a1}

primrec
  loc sGatewaySystem =
        {ch_i, ch_a}
  loc sGatewaySystemReq = {}
  loc sServiceCenter   = {}
  loc sGatewayReq   = {}
  loc sGateway = {ch_a1, ch_a2, ch_i1, ch_i2}
  loc sLoss   = {}
  loc sDelay = {}
  loc sSample = {}

primrec
  out sGatewaySystem =
        {ch_ack, ch_vc}
  out sGatewaySystemReq = {ch_ack, ch_vc}
  out sServiceCenter   = {ch_a}
  out sGatewayReq = {ch_ack, ch_vc, ch_i}
  out sGateway   = {ch_ack, ch_vc, ch_i}
  out sLoss   = {ch_i, ch_a2}
  out sDelay = {ch_a1, ch_i2}
  out sSample = {ch_ack, ch_vc, ch_i1}

Proofs for components

GatewaySystem:

lemma *spec_GatewaySystem1*:
correctInOutLoc sGatewaySystem
 by (*simp add: correctInOutLoc_def*)

lemma *spec_GatewaySystem2*:
correctComposition sGatewaySystem
 by (*simp add: correctComposition_def*)

lemma *spec_GatewaySystem3*:
correctCompositionIn sGatewaySystem
 by (*simp add: correctCompositionIn_def, auto*)

lemma *spec_GatewaySystem4*:
correctCompositionOut sGatewaySystem
 by (*simp add: correctCompositionOut_def, auto*)

lemma *spec_GatewaySystem5*:
correctCompositionLoc sGatewaySystem
 by (*simp add: correctCompositionLoc_def, auto*)

  GatewaySystemReq:

lemma *spec_GatewaySystemReq1*:
correctInOutLoc sGatewaySystemReq
 by (*simp add: correctInOutLoc_def*)

lemma *spec_GatewaySystemReq2*:
correctComposition sGatewaySystemReq
 by (*simp add: correctComposition_def*)

  ServiceCenter:

lemma *spec_ServiceCenter1*:
correctInOutLoc sServiceCenter
 by (*simp add: correctInOutLoc_def*)

lemma *spec_ServiceCenter2*:
correctComposition sServiceCenter
 by (*simp add: correctComposition_def*)

  GatewayReq:

lemma *spec_GatewayReq1*:
correctInOutLoc sGatewayReq
 by (*simp add: correctInOutLoc_def disjoint_def*)

lemma *spec_GatewayReq2*:
correctComposition sGatewayReq
 by (*simp add: correctComposition_def*)

Gateway:

lemma *spec_Gateway1*:
correctInOutLoc sGateway
  by (*simp add*: *correctInOutLoc_def disjoint_def*)

lemma *spec_Gateway2*:
correctComposition sGateway
  by (*simp add*: *correctComposition_def*)

lemma *spec_Gateway3*:
correctCompositionIn sGateway
  by (*simp add*: *correctCompositionIn_def disjoint_def*, *auto*)

lemma *spec_Gateway4*:
correctCompositionOut sGateway
  by (*simp add*: *correctCompositionOut_def disjoint_def*, *auto*)

lemma *spec_Gateway5*:
correctCompositionLoc sGateway
  by (*simp add*: *correctCompositionLoc_def*, *auto*)

Loss:

lemma *spec_Loss1*:
correctInOutLoc sLoss
  by (*simp add*: *correctInOutLoc_def disjoint_def*)

lemma *spec_Loss2*:
correctComposition sLoss
  by (*simp add*: *correctComposition_def*)

Delay:

lemma *spec_Delay1*:
correctInOutLoc sDelay
  by (*simp add*: *correctInOutLoc_def disjoint_def*)

lemma *spec_Delay2*:
correctComposition sDelay
  by (*simp add*: *correctComposition_def*)

Sample:

lemma *spec_Sample1*:
correctInOutLoc sSample
  by (*simp add*: *correctInOutLoc_def disjoint_def*)

lemma *spec_Sample2*:
correctComposition sSample
  by (*simp add*: *correctComposition_def*)
end

4.3.3. Gateway System: Architecture and Requirements

The FOCUS specification of the gateway system is presented below:

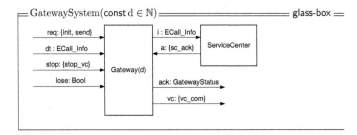

The stream *loss* is a time-synchronous one. It represents the connection status: the message **true** at the time interval t corresponds to the connection failure at this time interval, the message **false** at the time interval t means that at this time interval no data loss on the gateway connection.

The FOCUS representation of the specification *GatewaySystem* as plain text (constraint style):

$$\boxed{\begin{array}{l} \text{GatewaySystem(const d} \in \mathbb{N}) \hspace{4cm} \text{timed} \\[4pt] \begin{array}{ll} \text{in} & req : \{init, send\}; \quad dt : ECall\_Info; \\ & stop : stop\_vc; \quad lose : Bool \\ \text{out} & ack : GatewayStatus; \quad vc : \{vc\_com\} \\ \text{loc} & a : \{sc\_ack\}; \quad i : ECall\_Info \\ \end{array} \\[6pt] \hline \\ (ack, i, vc) := Gateway(d)(req, dt, a, stop, lose) \\ (a) := ServiceCenter(i) \end{array}}$$

We convert the FOCUS specification *GatewaySystem* into Isabelle/HOL predicate *GatewaySystem* schematically:

constdefs

$GatewaySystem ::$
$reqType\ istream \Rightarrow ECall\_Info\ istream \Rightarrow$
$stopType\ istream \Rightarrow bool\ istream \Rightarrow nat \Rightarrow$
$GatewayStatus\ istream \Rightarrow vcType\ istream \Rightarrow bool$
$GatewaySystem\ req\ dt\ stop\ lose\ d\ ack\ vc$
$\equiv$
$\exists\ a\ i.\ (Gateway\ req\ dt\ a\ stop\ lose\ d\ ack\ i\ vc) \wedge (ServiceCenter\ i\ a)$

The specification *GatewaySystemReq* specifies the requirements for the component *GatewaySystem*:

Assuming that the input streams *req* and *stop* can contain at every time interval at most one message, and assuming that the stream *lose* contains at every time interval exactly one message. If

✓ at any time interval t the gateway system is in the initial state, $\mathsf{ti}(ack, t) = \langle init\_state \rangle$, and

✓ at time interval $t+1$ the signal about crash comes at first time (more precise, the command to initiate the call to the Emergency Service Center), $\mathsf{ti}(req, t+1) = \langle init \rangle \ \land \ (\forall t_1 \in \mathbb{N}: \ t_1 \le t \rightarrow \mathsf{ti}(req, t1) = \langle \rangle)$, and

✓ after $3 + m$ time intervals the command to send the crash data comes at first time
$(\forall m \in \mathbb{N}: \ m \le k + 3 \rightarrow \mathsf{ti}(req, t+m) \ne \langle send \rangle) \ \land \ \mathsf{ti}(req, t+3+k) = \langle send \rangle$, and

✓ the gateway system has received until the time interval $t + 2$ the crash data, $\mathsf{last}^{ti}(dt, t+2) \ne \langle \rangle$

✓ there is no connection fails from the time interval t until the time interval $t + 4 + k + 2d$,
$\forall j \in \mathbb{N}: \ j \le (4 + k + d + d) \rightarrow \mathsf{ti}(lose, t+j) = \langle \mathsf{false} \rangle$

then at time interval $t + 4 + k + 2d$ the voice communication is established, $\mathsf{ti}(vc, t + 4 + k + d + d) = \langle vc\_com \rangle$.

```
═ GatewaySystemReq(const d ∈ ℕ) ══════════════════════ timed ═

  in    req : {init, send};  dt : ECall_Info;
        stop : stop_vc;  lose : Bool
  out   ack : GatewayStatus;  vc : {vc_com}
─────────────────────────────────────────────────────────────

  asm   msg₁(req) ∧ msg₁(stop) ∧ ts(lose)
  ─ ─ ─ ─ ─ ─ ─ ─ ─ ─ ─ ─ ─ ─ ─ ─ ─ ─ ─ ─ ─ ─ ─ ─ ─ ─ ─ ─ ─

  gar
  ∀ t, k ∈ ℕ :
    ti(ack, t) = ⟨init_state⟩ ∧ ti(req, t + 1) = ⟨init⟩
    ∧ ti(req, t + 2) = ⟨⟩
    ∧ (∀ t₁ ∈ ℕ :  t₁ ≤ t → ti(req, t1) = ⟨⟩)
    ∧ (∀ m ∈ ℕ :  m ≤ k + 3 → ti(req, t + m) ≠ ⟨send⟩)
    ∧ ti(req, t + 3 + k) = ⟨send⟩ ∧ last^ti(dt, t + 2) ≠ ⟨⟩
    ∧ (∀ j ∈ ℕ :  j ≤ (4 + k + d + d) → ti(lose, t + j) = ⟨false⟩)
    →
    ti(vc, t + 4 + k + d + d) = ⟨vc_com⟩
```

We convert the FOCUS specification *GatewaySystemReq* into Isabelle/HOL predicate *GatewaySystemReq* schematically:

constdefs

  GatewaySystemReq ::
    reqType istream $\Rightarrow$ *ECall_Info istream* $\Rightarrow$
    stopType istream $\Rightarrow$ *bool istream* $\Rightarrow$ *nat* $\Rightarrow$
    GatewayStatus istream $\Rightarrow$ *vcType istream* $\Rightarrow$ *bool*

  GatewaySystemReq req dt stop lose d ack vc

  $\equiv$

  $((msg\ (1::nat)\ req) \wedge (msg\ (1::nat)\ stop) \wedge (ts\ lose))$

  $\longrightarrow$

  $(\forall\ (t::nat)\ (k::nat).$
  $(\ ack\ t = [init\_state] \wedge req\ (Suc\ t) = [init]$
  $\wedge\ (\forall\ t1.\ t1 \leq t \longrightarrow req\ t1 = [])$
  $\wedge\ req\ (t{+}2) = []$
  $\wedge\ (\forall\ m.\ m < k + 3 \longrightarrow req\ (t + m) \neq [send])$
  $\wedge\ req\ (t{+}3{+}k)\ = [send] \wedge\ inf\_last\_ti\ dt\ (t{+}2) \neq []$
  $\wedge\ (\forall\ (j::nat).$
    $j \leq (4 + k + d + d) \longrightarrow lose\ (t{+}j) = [False])$
  $\longrightarrow vc\ (t + 4 + k + d + d) = [vc\_com])\ )$

4.3.4. ECall Service Center

The component *ServiceCenter* represents the behavior of the Emergency (ECall) Service Center from the gateway point of view: if at time t a message about a vehicle crash comes, it acknowledges this event by sending the at time $t + 1$ message *sc_ack* that represents the attempt to establish the voice communication with the driver or a passenger of the vehicle (*voice_communication* output message of the *Gateway* component) – if there is no connection failure, after d time intervals the voice communication will be started.

```
══ ServiceCenter ═══════════════════════════════════════ timed ══
  in     i : ECall_Info
  out    a : {sc_ack}
─────────────────────────────────────────────────────────────────
  ∀ t ∈ ℕ :
    ti(a, 0) = ⟨⟩
    ti(a, t + 1) = if ti(i, t) = ⟨⟩ then ⟨⟩ else ⟨sc_ack⟩ fi
```

The translation of this FOCUS specification into Isabelle/HOL predicate *ServiceCenter* is straightforward:

constdefs

  ServiceCenter ::
    ECall_Info istream $\Rightarrow$ *aType istream* $\Rightarrow$ *bool*
  ServiceCenter i a

  $\equiv$

  $\forall\ (t::nat).$
  $a\ 0 = [] \wedge\ a\ (Suc\ t) = (if\ (i\ t) = []\ then\ []\ else\ [sc\_ack])$

4.3.5. Gateway: Requirements Specification

We define the formal specification of the gateway requirement, presented in the previous section, as FOCUS specification *GatewayReq*:

1. If at time t the gateway is in the initial state *init_state*, and it gets the command to establish the connection with the central station, and also there is no environment connection problems during the next 2 time intervals, it establishes the connection at the time interval $t + 2$, $\mathsf{ti}(ack, t + 2) = \langle connection\_ok \rangle$.

2. If at time t the gateway has establish the connection, $\mathsf{ti}(ack, t) = \langle connection\_ok \rangle$, and it gets the command to send the E-Call data to the central station ($\mathsf{ti}(req, t + 1) = \langle send \rangle$), and also there is no environment connection problems during the next $d + 1$ time intervals, $\forall k \in \mathbb{N} : k \le d + 1 \rightarrow \mathsf{ti}(lose, t + k) = \langle \mathsf{false} \rangle$, then it sends the last corresponding data.[6] The central station becomes these date at the time $t + d$.

3. If the gateway becomes the acknowledgment from the central station that it has receives the sent E-Call data, and also there is no environment connection problems, then the voice communication is started.

The translation of this FOCUS specification into Isabelle/HOL predicate *GatewayReq* is straightforward.

GatewayReq(const d ∈ ℕ) ─────────────────────────── timed ══

in	$req : \{init\_connect\};\ dt : ECall\_Info;\ a : \{sc\_ack\};$
	$stop : stop\_vc;\ lose : Bool$
out	$ack : GatewayStatus;\ i : ECall\_Info;\ vc : \{voice\_com\}$

---

asm $\mathsf{msg}_1(req) \wedge \mathsf{msg}_1(a) \wedge \mathsf{msg}_1(stop) \wedge \mathsf{ts}(lose)$

- -

gar
$\forall\, t \in \mathbb{N} :$

   $\mathsf{ti}(ack, t) = \langle init\_state \rangle \ \wedge\ \mathsf{ti}(req, t + 1) = \langle init \rangle$
   $\wedge \mathsf{ti}(lose, t + 1) = \langle \mathsf{false} \rangle \ \wedge\ \mathsf{ti}(lose, t + 2) = \langle \mathsf{false} \rangle$
     $\rightarrow \mathsf{ti}(ack, t + 2) = \langle connection\_ok \rangle$

   $\mathsf{ti}(ack, t) = \langle connection\_ok \rangle \ \wedge\ \mathsf{ti}(req, t + 1) = \langle send \rangle$
   $\wedge (\forall k \in \mathbb{N} : k \le d + 1 \rightarrow \mathsf{ti}(lose, t + k) = \langle \mathsf{false} \rangle)$
     $\rightarrow \mathsf{ti}(i, t + d + 1) = \mathsf{last}^{ti}(dt, t) \ \wedge\ \mathsf{ti}(ack, t + 1) = \langle sending\_data \rangle$

   $\mathsf{ti}(ack, t + d) = \langle sending\_data \rangle \ \wedge\ \mathsf{ti}(a, t + 1) = \langle sc\_ack \rangle$
   $\wedge\ (\forall k \in \mathbb{N} : k \le d + 1 \rightarrow \mathsf{ti}(lose, t + k) = \langle \mathsf{false} \rangle)$
     $\rightarrow \mathsf{ti}(vc, t + d + 1) = \langle vc\_com \rangle$

---

[6]The FOCUS operator $\mathsf{last}^{ti}(s,t)$ returns the last nonempty time interval of the stream s until the tth time interval (see Section 2.5.13). If until the time t all intervals were empty, the empty message list is returned.

constdefs
  GatewayReq ::
    reqType istream $\Rightarrow$ *ECall_Info istream* $\Rightarrow$ *aType istream* $\Rightarrow$
    stopType istream $\Rightarrow$ *bool istream* $\Rightarrow$ *nat* $\Rightarrow$
    GatewayStatus istream $\Rightarrow$ *ECall_Info istream* $\Rightarrow$ *vcType istream*
    $\Rightarrow$ *bool*
  GatewayReq req dt a stop lose d ack i vc
    $\equiv$
  $((msg\ (1::nat)\ req) \wedge (msg\ (1::nat)\ a) \wedge$
  $(msg\ (1::nat)\ stop) \wedge (ts\ lose))$
    $\longrightarrow$
  $(\forall\ (t::nat).$
  $(\ ack\ t = [init\_state] \wedge req\ (Suc\ t) = [init] \wedge$
    $lose\ (t{+}1) = [False] \wedge lose\ (t{+}2) = [False]$
    $\longrightarrow ack\ (t{+}2) = [connection\_ok])$
  $\wedge$
  $(\ ack\ t = [connection\_ok] \wedge req\ (Suc\ t) = [send] \wedge$
    $(\forall\ (k::nat).\ k \leq (d{+}1) \longrightarrow lose\ (t{+}k) = [False])$
    $\longrightarrow i\ ((Suc\ t)\ +\ d) = inf\_last\_ti\ dt\ t$
      $\wedge\ ack\ (Suc\ t) = [sending\_data])$
  $\wedge$
  $(\ ack\ (t{+}d) = [sending\_data] \wedge a\ (Suc\ t) = [sc\_ack] \wedge$
    $(\forall\ (k::nat).\ k \leq (d{+}1) \longrightarrow lose\ (t{+}k) = [False])$
    $\longrightarrow vc\ ((Suc\ t)\ +\ d) = [vc\_com])\ )$

4.3.6. Gateway: Architecture Specification

The specification of the gateway architecture, *Gateway*, is parameterized one:
the paramether $d \in \mathbb{N}$ denotes the communication delay (between the central station and a vehicle). This component consists of three subcomponents:
Sample, *Delay*, and *Loss*.

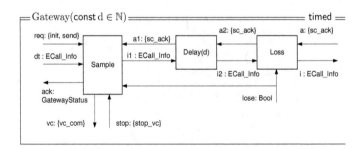

The FOCUS representation of the specification *Gateway* as plain text (constraint style):

```
┌═ Gateway(const d ∈ ℕ) ══════════════════════════════════ timed ══
│      req : {init, send};   dt : ECall_Info;   a : {sc_ack};
│  in
│      stop : stop_vc;   lose : Bool
│  out    ack : GatewayStatus;   i : ECall_Info;   vc : {vc_com}
│  loc    i1, i2 : ECall_Info;   a1, a2 : {sc_ack}
│─────────────────────────────────────────────────────────────────
│  (ack, i1, vc) := Sample(req, dt, a1, stop, lose)
│  (a1, i2) := Delay(d)(a2, i1)
│  (a2, i) := Loss(lose, a, i2)
```

We convert the FOCUS specification *Gateway* into Isabelle/HOL predicate *Gateway* schematically:

constdefs
$Gateway$ ::
   $reqType\ istream \Rightarrow ECall\_Info\ istream \Rightarrow aType\ istream \Rightarrow$
   $stopType\ istream \Rightarrow bool\ istream \Rightarrow nat \Rightarrow$
   $GatewayStatus\ istream \Rightarrow ECall\_Info\ istream \Rightarrow vcType\ istream \Rightarrow bool$
$Gateway\ req\ dt\ a\ stop\ lose\ d\ ack\ i\ vc$
  $\equiv \exists\ i1\ i2\ a1\ a2.$
  $(Sample\ req\ dt\ a1\ stop\ lose\ ack\ i1\ vc) \wedge$
  $(Delay\ a2\ i1\ d\ a1\ i2) \wedge$
  $(Loss\ lose\ a\ i2\ a2\ i)$

4.3.7. Sample Component

The component *Sample* represents the logic of the gateway component. If it receives from a ECall application of a vehicle the command to initiate the call to the Emergency Service Center it tries to establish the connection. If the connection is established, and the component *Sample* receives from a ECall application of a vehicle the command to send the crash data, which were already received and stored in the internal buffer of the gateway, these data will be resent to the Emergency Service Center. After that this component waits to the acknowledgment from the Emergency Service Center. If the acknowledgment is received, the voice communication will be established, assuming that there is no connection fails.

We present here two variants of the FOCUS specification of the component *Sample*: using a new variant of the FOCUS tables, tiTable, and using the plain text FOCUS notation. These two specification kinds are semantically equivalent according the definition of the tiTable.

For the component *Sample* we have the assumption, that the streams *req*, *a1*, and *stop* can contain at every time interval at most one message, and also that the stream *loss* must contain at every time interval exactly one message. This component uses local variables *st* and *buffer* (more precisely, a local variable *buffer* and a state variable *st*). The guarantee part of the component *Sample* consists of an externally defined tiTable *SampleT* (see below) and an expression which says how the local variable *buffer* is computed.

```
══ Sample ══════════════════════════════════════════ timed ══
      req : {init, send};  dt : ECall_Info;  a1 : {sc_ack};
  in  stop : {stop_vc};  lose : Bool
  out ack : GatewayStatus;  i1 : ECall_Info;  vc : {voice_com}
─────────────────────────────────────────────────────────
  local  st : GatewayStatus;  buffer : ECall_Info*
- - - - - - - - - - - - - - - - - - - - - - - - - - - - - - - -
  init   st = init_state;  buffer = ⟨⟩;
- - - - - - - - - - - - - - - - - - - - - - - - - - - - - - - -
  asm    msg₁(req) ∧ msg₁(a1) ∧ msg₁(stop) ∧ ts(lose)
- - - - - - - - - - - - - - - - - - - - - - - - - - - - - - - -
  gar
     ∀ t ∈ ℕ :
        buffer′ = if ti(dt, t) = ⟨⟩ then buffer else ti(dt, t) fi

     tiTable SampleT
```

Towards Section 2.7, the representation of the component *Sample* in Isabelle/HOL must consist of three predicates: *tiTable_SampleT* (Isabelle/HOL representation of the corresponding tiTable), *Sample_L* (an auxiliary predicate to represent the local and the state variables in Isabelle/HOL), and the "main" predicate *Sample*. The definition of the predicates *Sample* and *Sample_L* can be done straightforward according to the translation schema (see Sections 2.14 and 2.7).

constdefs
Sample ::
reqType istream ⇒ *ECall_Info istream* ⇒ *aType istream* ⇒
stopType istream ⇒ *bool istream* ⇒
GatewayStatus istream ⇒ *ECall_Info istream* ⇒ *vcType istream*
⇒ *bool*
Sample req dt a1 stop lose ack i1 vc
  ≡
((*msg* (1::*nat*) *req*) ∧
 (*msg* (1::*nat*) *a1*) ∧
 (*msg* (1::*nat*) *stop*))
⟶
(∃ *st buffer*.
(*Sample_L req dt a1 stop lose*
        (*fin_inf_append* [*init_state*] *st*)
        (*fin_inf_append* [[]] *buffer*)
        ack i1 vc st buffer))

tiTable SampleT (univ r : {$init, send$}*; x : {sc_ack}*; y : {$stop\_vc$}*; z : Bool*): $\forall t \in \mathbb{N}$

	req	a1	stop	lose	ack	i1	vc	st'	Assumption
1	$\langle init \rangle$	x	y	z	$\langle call \rangle$	$\langle \rangle$	$\langle \rangle$	$call$	$st = init\_state$
2	r	x	y	z	$\langle init\_state \rangle$	$\langle \rangle$	$\langle \rangle$	$init\_state$	$st = init\_state \wedge r \neq \langle init \rangle$
3	r	x	y	$\langle false \rangle$	$\langle connection\_ok \rangle$	$\langle \rangle$	$\langle \rangle$	$connection\_ok$	$st = call \vee (st = connection\_ok \wedge r \neq \langle send \rangle)$
4	r	x	y	$\langle true \rangle$	$\langle init\_state \rangle$	$\langle \rangle$	$\langle \rangle$	$init\_state$	$st = call \vee st = connection\_ok \vee st = sending\_data$
5	$\langle send \rangle$	x	y	$\langle false \rangle$	$\langle sending\_data \rangle$	$buffer$	$\langle \rangle$	$sending\_data$	$st = connection\_ok$
6	r	$\langle \rangle$	y	$\langle false \rangle$	$\langle sending\_data \rangle$	$\langle \rangle$	$\langle \rangle$	$sending\_data$	$st = sending\_data$
7	r	$\langle sc\_ack \rangle$	y	$\langle false \rangle$	$\langle voice\_com \rangle$	$\langle \rangle$	$\langle vc\_com \rangle$	$voice\_com$	$st = sending\_data$
8	r	x	$\langle \rangle$	$\langle false \rangle$	$\langle voice\_com \rangle$	$\langle \rangle$	$\langle vc\_com \rangle$	$voice\_com$	$st = voice\_com$
9	r	x	$\langle \rangle$	$\langle true \rangle$	$\langle voice\_com \rangle$	$\langle \rangle$	$\langle \rangle$	$voice\_com$	$st = voice\_com$
10	r	x	$\langle stop\_vc \rangle$	z	$\langle init\_state \rangle$	$\langle \rangle$	$\langle \rangle$	$init\_state$	$st = voice\_com$

constdefs
Sample_L ::
reqType istream ⇒ *ECall_Info istream* ⇒ *aType istream* ⇒
stopType istream ⇒ *bool istream* ⇒
(*nat* ⇒ *GatewayStatus*) ⇒ (*nat* ⇒ *ECall_Info list*) ⇒
GatewayStatus istream ⇒ *ECall_Info istream* ⇒ *vcType istream*
⇒ (*nat* ⇒ *GatewayStatus*) ⇒ (*nat* ⇒ *ECall_Info list*) ⇒ *bool*
Sample_L req dt a1 stop lose st_in buffer_in
       ack i1 vc st_out buffer_out
  ≡
(∀ (*t::nat*).
  buffer_out t = (*if dt t* = [] *then buffer_in t else dt t*))
  ∧
(*tiTable_SampleT req a1 stop lose st_in buffer_in ack i1 vc st_out*)

The FOCUS predicate *tiTable_SampleT* is the plain text of the tiTable *SampleT*. The schematic translation of its FOCUS specification into Isabelle/HOL predicate *tiTable_SampleT* is trivial (see Section B.5).

⚠ **Remark:** In this tiTable we have used universally quantified variables, r, x, y, and z. We can omit the input stream cells with such an variable, if this variable does not to be used in this line neither in the assumption nor to define the output (stream) cells. As shown in the predicate specification *tiTable_SampleT*, we can also replace a variable by the corresponding expression with the time interval operator, for example, the variable r in the cells of the stream *req* corresponds to the expression $ti(req, t)$. But we also can use more direct, non-simplified, representation, as shown in the predicate specification *tiTable_SampleT_ext* (see below). The schematic translation of its FOCUS specification into Isabelle/HOL predicate *tiTable_SampleT_ext* is also trivial (see Section B.5). The equivalence of these representation is proved by a lemma *univ_tiTable_Sample* (see Section B.7).

We can also represent the logic of this component as a timed state transition diagram (TSTD, see Section 2.8). The TSTD from Fig.4.1 is semantically equivalent to the following tiTable *SampleT_ext*, which is an "extraction" of the tiTable *SampleT*: lines of the tiTable *SampleT*, which assumption cells contain disjunctions, are "split" into x lines in the "extracted" tiTable, where x is the number of elements in the disjunction. Thus, if we compare the tables *SampleT* and *SampleT_ext*, we can easily see, the following relation between their lines:

SampleT	1	2	3	4	5	6	7	8	9	10
SampleT_ext	1	2	3, 4	5, 6, 7	8	9	10	11	12	13

___tiTable_SampleT_____

$req \in \{init, send\}^{\infty};\quad dt \in ECall\_Info^{\infty};\quad a1 \in \{sc\_ack\}^{\infty};$
$stop \in \{stop\_vc\}^{\infty};\quad lose \in \mathbb{Bool}^{\infty};$
$st \in GatewayStatus;\quad buffer \in ECall\_Info^{*};$
$ack \in GatewayStatus^{\infty};\quad i1 \in ECall\_Info^{\infty};\quad vc \in \{vc\_com\}^{\infty};$

$\forall\, t \in \mathbb{N}:$

$\quad st = init\_state \wedge \mathrm{ti}(req, t) = \langle init \rangle$
$\qquad \Rightarrow \mathrm{ti}(ack, t) = \langle call \rangle \wedge \mathrm{ti}(i1, t) = \langle\rangle \wedge \mathrm{ti}(vc, t) = \langle\rangle \wedge st' = call$

$\quad st = init\_state \wedge \mathrm{ti}(req, t) \neq \langle init \rangle$
$\qquad \Rightarrow \mathrm{ti}(ack, t) = \langle init\_state \rangle \wedge \mathrm{ti}(i1, t) = \langle\rangle \wedge \mathrm{ti}(vc, t) = \langle\rangle \wedge st' = init\_state$

$\quad (st = call \vee (st = connection\_ok \wedge \mathrm{ti}(req, t) \neq \langle send \rangle)) \wedge \mathrm{ti}(lose, t) = \langle \mathsf{false} \rangle$
$\qquad \Rightarrow \mathrm{ti}(ack, t) = \langle connection\_ok \rangle \wedge \mathrm{ti}(i1, t) = \langle\rangle \wedge$
$\qquad\quad \mathrm{ti}(vc, t) = \langle\rangle \wedge st' = connection\_ok$

$\quad (st = call \vee st = connection\_ok \vee st = sending\_data) \wedge \mathrm{ti}(lose, t) = \langle \mathsf{true} \rangle$
$\qquad \Rightarrow \mathrm{ti}(ack, t) = \langle init\_state \rangle \wedge \mathrm{ti}(i1, t) = \langle\rangle \wedge$
$\qquad\quad \mathrm{ti}(vc, t) = \langle\rangle \wedge st' = init\_state$

$\quad st = connection\_ok \wedge \mathrm{ti}(lose, t) = \langle \mathsf{false} \rangle \wedge \mathrm{ti}(req, t) = \langle send \rangle$
$\qquad \Rightarrow \mathrm{ti}(ack, t) = \langle sending\_data \rangle \wedge \mathrm{ti}(i1, t) = buffer \wedge$
$\qquad\quad \mathrm{ti}(vc, t) = \langle\rangle \wedge st' = sending\_data$

$\quad st = sending\_data \wedge \mathrm{ti}(lose, t) = \langle \mathsf{false} \rangle \wedge \mathrm{ti}(a1, t) = \langle\rangle$
$\qquad \Rightarrow \mathrm{ti}(ack, t) = \langle sending\_data \rangle \wedge \mathrm{ti}(i1, t) = \langle\rangle \wedge$
$\qquad\quad \mathrm{ti}(vc, t) = \langle\rangle \wedge st' = sending\_data$

$\quad st = sending\_data \wedge \mathrm{ti}(lose, t) = \langle \mathsf{false} \rangle \wedge \mathrm{ti}(a1, t) = \langle sc\_ack \rangle$
$\qquad \Rightarrow \mathrm{ti}(ack, t) = \langle voice\_com \rangle \wedge \mathrm{ti}(i1, t) = \langle\rangle \wedge$
$\qquad\quad \mathrm{ti}(vc, t) = \langle vc\_com \rangle \wedge st' = voice\_com$

$\quad st = voice\_com \wedge \mathrm{ti}(lose, t) = \langle \mathsf{false} \rangle \wedge \mathrm{ti}(stop, t) = \langle\rangle$
$\qquad \Rightarrow \mathrm{ti}(ack, t) = \langle voice\_com \rangle \wedge \mathrm{ti}(i1, t) = \langle\rangle \wedge$
$\qquad\quad \mathrm{ti}(vc, t) = \langle vc\_com \rangle \wedge st' = \langle voice\_com \rangle$

$\quad st = voice\_com \wedge \mathrm{ti}(lose, t) = \langle \mathsf{true} \rangle \wedge \mathrm{ti}(stop, t) = \langle\rangle$
$\qquad \Rightarrow \mathrm{ti}(ack, t) = \langle voice\_com \rangle \wedge \mathrm{ti}(i1, t) = \langle\rangle \wedge$
$\qquad\quad \mathrm{ti}(vc, t) = \langle\rangle \wedge st' = \langle voice\_com \rangle$

$\quad st = voice\_com \wedge \mathrm{ti}(stop, t) = \langle stop\_vc \rangle$
$\qquad \Rightarrow \mathrm{ti}(ack, t) = \langle init\_state \rangle \wedge \mathrm{ti}(i1, t) = \langle\rangle \wedge$
$\qquad\quad \mathrm{ti}(vc, t) = \langle\rangle \wedge st' = \langle init\_state \rangle$

_tiTable_SampleT_ext_

$req \in \{init, send\}^{\infty}; \quad dt \in ECall\_Info^{\infty}; \quad a1 \in \{sc\_ack\}^{\infty};$
$stop \in \{stop\_vc\}^{\infty}; \quad lose \in \mathbb{Bool}^{\infty};$
$st \in GatewayStatus; \quad buffer \in ECall\_Info^{*};$
$ack \in GatewayStatus^{\infty}; \quad i1 \in ECall\_Info^{\infty}; \quad vc \in \{vc\_com\}^{\infty};$

$\forall t \in \mathbb{N}, \; r \in \{init, send\}^{*}, \; x \in \{sc\_ack\}^{*}, \; y \in \{stop\_vc\}^{*}, \; z \in \mathbb{Bool}^{*}:$

$st = init\_state \wedge \mathsf{ti}(req, t) = \langle init \rangle \wedge$
$\mathsf{ti}(a1, t) = x \wedge \mathsf{ti}(stop, t) = y \wedge \mathsf{ti}(lose, t) = z \wedge$
$\quad\quad \Rightarrow \mathsf{ti}(ack, t) = \langle call \rangle \wedge \mathsf{ti}(i1, t) = \langle \rangle \wedge \mathsf{ti}(vc, t) = \langle \rangle \wedge st' = call$

$st = init\_state \wedge r \neq \langle init \rangle \wedge$
$\mathsf{ti}(req, t) = r \wedge \mathsf{ti}(a1, t) = x \wedge \mathsf{ti}(stop, t) = y \wedge \mathsf{ti}(lose, t) = z$
$\quad\quad \Rightarrow \mathsf{ti}(ack, t) = \langle init\_state \rangle \wedge \mathsf{ti}(i1, t) = \langle \rangle \wedge$
$\quad\quad\quad \mathsf{ti}(vc, t) = \langle \rangle \wedge st' = init\_state$

$(st = call \vee (st = connection\_ok \wedge r \neq \langle send \rangle)) \wedge$
$\mathsf{ti}(req, t) = r \wedge \mathsf{ti}(a1, t) = x \wedge \mathsf{ti}(stop, t) = y \wedge \mathsf{ti}(lose, t) = \langle \mathsf{false} \rangle$
$\quad\quad \Rightarrow \mathsf{ti}(ack, t) = \langle connection\_ok \rangle \wedge$
$\quad\quad\quad \mathsf{ti}(i1, t) = \langle \rangle \wedge \mathsf{ti}(vc, t) = \langle \rangle \wedge st' = connection\_ok$

$(st = call \vee st = connection\_ok \vee st = sending\_data) \wedge \mathsf{ti}(req, t) = r \wedge$
$\mathsf{ti}(a1, t) = x \wedge \mathsf{ti}(stop, t) = y \wedge \mathsf{ti}(lose, t) = \langle \mathsf{true} \rangle$
$\quad\quad \Rightarrow \mathsf{ti}(ack, t) = \langle init\_state \rangle \wedge \mathsf{ti}(i1, t) = \langle \rangle \wedge$
$\quad\quad\quad \mathsf{ti}(vc, t) = \langle \rangle \wedge st' = init\_state$

$st = connection\_ok \wedge \mathsf{ti}(req, t) = \langle send \rangle \wedge$
$\mathsf{ti}(a1, t) = x \wedge \mathsf{ti}(stop, t) = y \wedge \mathsf{ti}(lose, t) = \langle \mathsf{false} \rangle$
$\quad\quad \Rightarrow \mathsf{ti}(ack, t) = \langle sending\_data \rangle$
$\quad\quad\quad \wedge \mathsf{ti}(i1, t) = buffer \wedge \mathsf{ti}(vc, t) = \langle \rangle \wedge st' = sending\_data$

$st = sending\_data \wedge \mathsf{ti}(req, t) = r \wedge$
$\mathsf{ti}(a1, t) = \langle \rangle \wedge \mathsf{ti}(stop, t) = y \wedge \mathsf{ti}(lose, t) = \langle \mathsf{false} \rangle$
$\quad\quad \Rightarrow \mathsf{ti}(ack, t) = \langle sending\_data \rangle \wedge \mathsf{ti}(i1, t) = \langle \rangle \wedge$
$\quad\quad\quad \mathsf{ti}(vc, t) = \langle \rangle \wedge st' = sending\_data$

$st = sending\_data \wedge \mathsf{ti}(req, t) = r \wedge$
$\mathsf{ti}(a1, t) = \langle sc\_ack \rangle \wedge \mathsf{ti}(stop, t) = y \wedge \mathsf{ti}(lose, t) = \langle \mathsf{false} \rangle$
$\quad\quad \Rightarrow \mathsf{ti}(ack, t) = \langle voice\_com \rangle \wedge \mathsf{ti}(i1, t) = \langle \rangle \wedge$
$\quad\quad\quad \mathsf{ti}(vc, t) = \langle vc\_com \rangle \wedge st' = voice\_com$

$st = voice\_com \wedge \mathsf{ti}(req, t) = r \wedge$
$\mathsf{ti}(a1, t) = x \wedge \mathsf{ti}(stop, t) = \langle \rangle \wedge \mathsf{ti}(lose, t) = \langle \mathsf{false} \rangle$
$\quad\quad \Rightarrow \mathsf{ti}(ack, t) = \langle voice\_com \rangle \wedge \mathsf{ti}(i1, t) = \langle \rangle \wedge$
$\quad\quad\quad \mathsf{ti}(vc, t) = \langle vc\_com \rangle \wedge st' = \langle voice\_com \rangle$

$st = voice\_com \wedge \mathsf{ti}(req, t) = r \wedge$
$\mathsf{ti}(a1, t) = x \wedge \mathsf{ti}(stop, t) = \langle \rangle \wedge \mathsf{ti}(lose, t) = \langle \mathsf{true} \rangle$
$\quad\quad \Rightarrow \mathsf{ti}(ack, t) = \langle voice\_com \rangle \wedge \mathsf{ti}(i1, t) = \langle \rangle \wedge$
$\quad\quad\quad \mathsf{ti}(vc, t) = \langle \rangle \wedge st' = \langle voice\_com \rangle$

$st = voice\_com \wedge \mathsf{ti}(req, t) = r \wedge$
$\mathsf{ti}(a1, t) = x \wedge \mathsf{ti}(stop, t) = \langle stop\_vc \rangle \wedge \mathsf{ti}(lose, t) = z$
$\quad\quad \Rightarrow \mathsf{ti}(ack, t) = \langle init\_state \rangle \wedge \mathsf{ti}(i1, t) = \langle \rangle \wedge$
$\quad\quad\quad \mathsf{ti}(vc, t) = \langle \rangle \wedge st' = \langle init\_state \rangle$

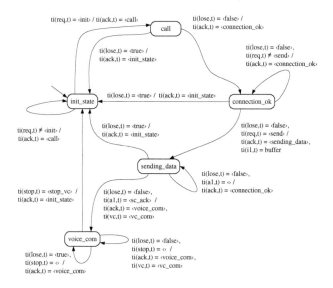

Figure 4.1.: Timed state transition diagram for the component Sample

tiTable SampleT_ext (univ $r : \{init, send\}^*$; $x : \{sc\_ack\}^*$; $y : \{stop\_vc\}^*$; $z : \mathbb{B}ool^*$): $\forall t \in \mathbb{N}$

	req	a1	stop	lose	ack	i1	vc	st'	Assumption
1	⟨init⟩	x	y	z	⟨call⟩	◇	◇	call	$st = init\_state$
2	r	x	y	z	⟨init_state⟩	◇	◇	init_state	$st = init\_state \wedge r \neq \langle init \rangle$
3	r	x	y	⟨false⟩	⟨connection_ok⟩	◇	◇	connection_ok	$st = call$
4	r	x	y	⟨false⟩	⟨connection_ok⟩	◇	◇	connection_ok	$st = connection\_ok \wedge r \neq \langle send \rangle$
5	r	x	y	⟨true⟩	⟨init_state⟩	◇	◇	init_state	$st = call$
6	r	x	y	⟨true⟩	⟨init_state⟩	◇	◇	init_state	$st = connection\_ok$
7	r	x	y	⟨true⟩	⟨init_state⟩	◇	◇	init_state	$st = sending\_data$
8	⟨send⟩	x	y	⟨false⟩	⟨sending_data⟩	buffer	◇	sending_data	$st = connection\_ok$
9	r	◇	y	⟨false⟩	⟨sending_data⟩	◇	◇	sending_data	$st = sending\_data$
10	r	⟨sc_ack⟩	y	⟨false⟩	⟨voice_com⟩	◇	⟨vc_com⟩	voice_com	$st = sending\_data$
11	r	x	◇	⟨false⟩	⟨voice_com⟩	◇	⟨vc_com⟩	voice_com	$st = voice\_com$
12	r	x	◇	⟨true⟩	⟨voice_com⟩	◇	◇	voice_com	$st = voice\_com$
13	r	x	⟨stop_vc⟩	z	⟨init_state⟩	◇	◇	init_state	$st = voice\_com$

4.3.8. Delay Component

The component *Delay* models the communication delay. Its specification is parameterized one: it inherits the parameter of the component *Gateway*. This component simply delays all input messages on d time intervals. During the first d time intervals no output message will be produced. We convert this FOCUS specification into Isabelle/HOL predicate *Delay* schematically.

```
═ Delay(const d ∈ ℕ) ═════════════════════════════════ timed ═
  in      a2 : {sc_ack};   i1 : ECall_Info
  out     a1 : {sc_ack};   i2 : ECall_Info
─────────────────────────────────────────────────────────────
  ∀ t ∈ ℕ :
    t < d  →  ti(a1, t) = ⟨⟩ ∧ ti(i2, t) = ⟨⟩
    ti(a1, t + d) = ti(a2, t)
    ti(i2, t + d) = ti(i1, t)
```

constdefs
Delay ::
  aType istream ⇒ *ECall_Info istream* ⇒ *nat* ⇒
  aType istream ⇒ *ECall_Info istream* ⇒ *bool*
Delay a2 i1 d a1 i2
  ≡
  ∀ (*t::nat*).
  (*t* < *d* ⟶ *a1 t* = [] ∧ *i2 t* = []) ∧
  (*a1* (*t+d*) = *a2 t*) ∧ (*i2* (*t+d*) = *i1 t*)

4.3.9. Loss Component

The component *Loss* models the communication loss between the central station and the vehicle gateway: if during time interval t from the component *Loss_Oracle* no message about a lost connection comes, $ti(lose,t) = \langle\text{false}\rangle$, the messages come during time interval t via the input channels a and $i2$ will be forwarded without any delay via channels $a2$ and i respectivelly. Otherwiese all messages come during time interval t will be lost.

```
═ Loss ═══════════════════════════════════════════════ timed ═
  in      lose : Bool;   a : {sc_ack};   i2 : ECall_Info
  out     a2 : {sc_ack};   i : ECall_Info
─────────────────────────────────────────────────────────────
  ∀ t ∈ ℕ :
    if ti(lose, t) = ⟨false⟩
    then ti(a2, t) = ti(a, t) ∧ ti(i, t) = ti(i2, t)
    else ti(a2, t) = ⟨⟩ ∧ ti(i, t) = ⟨⟩
    fi
```

We convert the FOCUS specification *Loss* into Isabelle/HOL predicate *Loss* schematically:

constdefs

Loss ::

  bool istream $\Rightarrow$ *aType istream* $\Rightarrow$ *ECall_Info istream* $\Rightarrow$

  aType istream $\Rightarrow$ *ECall_Info istream* $\Rightarrow$ *bool*

Loss lose a i2 a2 i

  $\equiv$

$\forall$ *(t::nat).*

(*if lose t =* [*False*]

  then a2 t = a t $\wedge$ *i t = i2 t*

  else a2 t = [] $\wedge$ *i t =* [])

4.3.10. Verification of the Gateway

To show that the specified gateway architecture fulfills the requirements we need to show that the specification *Gateway* is a refinement of the specification *GatewayReq*. Therefore, we need to define and to prove the following lemma:

lemma *Gateway_L0*:

Gateway req dt a stop lose d ack i vc

$\Longrightarrow$

GatewayReq req dt a stop lose d ack i vc

To prove this lemma we used first of all the definition of the predicate *GatewayReq*, use the Isabelle/HOL automatic proof strategies, and apply the following lemmas: lemma *Gateway_L1* represents the first requirement from the specification *GatewayReq* (see Section 4.3.5), lemmas *Gateway_L2* and *Gateway_L3* represent two conjuncts of the second requirement from the specification *GatewayReq*, and lemma *Gateway_L4* corresponds to the third requirement from the specification *GatewayReq*. The Isabelle/HOL proofs of all these lemmas are given in Appendix Section B.7.

✓

    lemma *Gateway_L1*:

    [*Gateway req dt a stop lose d ack i vc*;

      msg (Suc 0) req; *msg (Suc 0) a*; *msg (Suc 0) stop*;

      ack t = [*init_state*]; *req (Suc t) =* [*init*]; *ts lose*;

      lose (Suc t) = [*False*]; *lose (Suc (Suc t)) =* [*False*]]

    $\Longrightarrow$ *ack (Suc (Suc t)) =* [*connection_ok*]

    Assuming that for the stream *lose* the predicate *ts* holds, and assuming also that the streams *req*, *a*, *stop* have at every time interval at most one message (*msg (Suc 0) req*, *msg (Suc 0) a*, *msg (Suc 0) stop*). If at time *t* the gateway is in the initial state, *ack t =* [*init_state*], it gets the command to establish the connection with the central station, *req (Suc t) =* [*init*], and also there is no environment connection problems during the next 2 time intervals, (*lose (Suc t) =* [*False*] and *lose (Suc (Suc t)) =* [*False*], thet it establishes the connection at the time interval *t* + 2: *ack (Suc (Suc t)) =* [*connection_ok*].

✓

lemma *Gateway_L2*:

⟦ *Gateway req dt a stop lose i ack i vc*;
  msg (Suc 0) req; *msg (Suc 0) a*; *msg (Suc 0) stop*;
  ack t = [*connection_ok*]; *req (Suc t)* = [*send*];
  ∀ *k*≤*Suc d. lose (t + k)* = [*False*]; *ts lose* ⟧
  ⟹ *i (Suc (t + d))* = *inf_last_ti dt t*

Assuming that for the stream *lose* the predicate *ts* holds, and assuming also that the streams *req, a, stop* have at every time interval at most one message (*msg (Suc 0) req, msg (Suc 0) a, msg (Suc 0) stop*). If at time *t* the gateway is in the state where the connection is established, *ack t* = [*connection_ok*], and it gets the command to send the crash data to the central station, *req (Suc t)* = [*send*], and also there is no environment connection problems during the next *d* + 1 time intervals[7], ∀*k* ≤ *Suc d. lose (t + k)* = [*False*], then it sends the last crash data, which it has received via the channel *dt* before the time *t*, at the time interval *t* + *d* + 1: *i (Suc (t + d))* = *inf_last_ti dt t*.

✓

lemma *Gateway_L3*:

⟦ *Gateway req dt a stop lose d ack i vc*;
  msg (Suc 0) req; *msg (Suc 0) a*; *msg (Suc 0) stop*;
  ts lose; *ack t* = [*connection_ok*];
  req (Suc t) = [*send*]; ∀ *k*≤*Suc d. lose (t + k)* = [*False*]⟧
  ⟹ *ack (Suc t)* = [*sending_data*]

Assuming that for the stream *lose* the predicate *ts* holds, and assuming also that the streams *req, a, stop* have at every time interval at most one message (*msg (Suc 0) req, msg (Suc 0) a, msg (Suc 0) stop*). If at time *t* the gateway is in the state where the connection is established, *ack t* = [*connection_ok*], and it gets the command to send the crash data to the central station, *req (Suc t)* = [*send*], and also there is no environment connection problems during the next *d*+1 time intervals (∀*k* ≤ *Suc d. lose (t + k)* = [*False*]), then it sends the crash data and goes in the corresponding control state: *ack (Suc t)* = [*sending_data*].

✓

lemma *Gateway_L4*:

⟦ *Gateway req dt a stop lose d ack i vc*;
  msg (Suc 0) req; *msg (Suc 0) a*; *msg (Suc 0) stop*;
  ts lose; *ack (t + d)* = [*sending_data*]; *a (Suc t)* = [*sc_ack*];
  ∀ *k*≤*Suc d. lose (t + k)* = [*False*]⟧
  ⟹ *vc (Suc (t + d))* = [*vc_com*]

Assuming that for the stream *lose* the predicate *ts* holds, and assuming also that the streams *req, a, stop* have at every time interval at most one message (*msg (Suc 0) req, msg (Suc 0) a, msg (Suc 0) stop*).

---

[7]*d* is the delay of trasnsmission between the gateway and the central station.

If at time $t + d$ the gateway is in the state where the crash data were
sent, *ack (t+d) = [sending_data]*, and the central station sends at time $t+1$
the acknowledgment that this data were received, *a (Suc t) = [sc_ack]*, and
also there is no environment connection problems during the $d + 1$ time
intervals from the time t, $\forall k \leq Suc\ d.\ lose\ (t + k) = [False]$, then the voice
connection is established at time interval $t + d + 1$:
vc (Suc (t+d)) = [vc_com].

4.3.11. Verification of the Gateway System

To show that the specified gateway architecture fulfills the requirements we need
to show that the specification *GatewaySystem* is a refinement of the specification
GatewaySystemReq. Therefore, we need to define and to prove the following
lemma:

lemma *GatewaySystem_L0*:
GatewaySystem req dt stop lose d ack vc
$\implies$
GatewaySystemReq req dt stop lose d ack vc

To prove this lemma we used first of all the definitions of the predicates *Gateway-
SystemReq* and *GatewaySystem*, and clarify the resulting goal. After that we add
two new assumptions to the goal:

- ✓ The stream a has at very time interval at most one message (this property
  of the central station component is proved as lemma *ServiceCenter_a_msg*,
  see Section B.7).[8] This assumption is necessary as one of the gateway
  assumptions about the environment.

- ✓ The predicate *GatewayReq* holds for the corresponding streams, i.e. that
  the gateway fulfills its requirements (according lemma *Gateway_L0*, see
  Sections 4.3.10 and B.7).

  This assumption is needed to simplify the proof – now we can prove a
  number of system properties directly from the *properties* of the gateway,
  without extraction the definitions of the gateway architecture and the
  properties of its components.

Now we split the goal into 4 subgoals (these 4 cases are needed because of
the definition of the predicate *inf_last_ti*): $\mathsf{ti}(dt, t + 1) = \langle \rangle \wedge \mathsf{ti}(dt, t + 2) = \langle \rangle$,
$\mathsf{ti}(dt, t + 1) = \langle \rangle \wedge \mathsf{ti}(dt, t + 2) \neq \langle \rangle$, $\mathsf{ti}(dt, t + 1) \neq \langle \rangle \wedge \mathsf{ti}(dt, t + 2) = \langle \rangle$,
and $\mathsf{ti}(dt, t + 1) \neq \langle \rangle \wedge \mathsf{ti}(dt, t + 2) \neq \langle \rangle$. We solve these subgoals using the
Isabelle/HOL automatic proof strategies as well as lemmas *GatewaySystem_L2*
(for the 1. case) and *GatewaySystem_L3* (for the cases 2–4).

The lemma *GatewaySystem_L2* says: the voice communication must be established at the time interval $2 * d + t + 4 + k$, if

---

[8]This stream goes from the central station to the gateway, see Section 4.3.3.

✓ The predicate *ts* holds for the stream *lose*, and the streams *req*, *a*, *stop* have at every time interval at most one message;

✓ The predicates *Gateway*, *GatewayReq*, and *ServiceCenter* are true for the corresponding streams;

✓ At time t the gateway is in the initial state ($ack\ t = [init\_state]$), and it gets the command to establish the connection with the central station ($req\ (Suc\ t) = [init]$), besides there was no such command before the time t ($\forall t1 < t.\ req\ t1 = []$);

✓ After at least 2 time intervals the gateway get (ones) the command to send the crash data to the central station ($req\ (t + 3 + k) = [send]$, $\forall m \leq k + 2.\ req\ (t + m) \neq [send]$);

✓ The gateway has received until the time t some crash data ($inf\_last\_ti\ dt\ t \neq []$);

✓ There is no environment connection problems during the next $2 * d + 4 + k$ time intervals, where d is the communication delay between the gateway and the central station.

lemma *GatewaySystem_L2*:
$[Gateway\ req\ dt\ a\ stop\ lose\ d\ ack\ i\ vc;\ ServiceCenter\ i\ a;$
$GatewayReq\ req\ dt\ a\ stop\ lose\ d\ ack\ i\ vc;$
$msg\ (Suc\ 0)\ req;\ msg\ (Suc\ 0)\ stop;\ ts\ lose;\ msg\ (Suc\ 0)\ a;$
$ack\ t = [init\_state];\ req\ (Suc\ t) = [init];$
$\forall t1 \leq t.\ req\ t1 = [];$
$\forall m \leq k + 2.\ req\ (t + m) \neq [send];\ req\ (t + 3 + k) = [send];$
$inf\_last\_ti\ dt\ t \neq [];\ \forall j \leq 2 * d + (4 + k).\ lose\ (t + j) = [False]]$
$\implies vc\ (2 * d + (t + (4 + k))) = [vc\_com]$

To prove this lemma we add first of all the following assumptions:

(1) The gateway establishes the connection at the time interval $t + 2$, $ack\ (Suc\ (Suc\ t)) = [connection\_ok]$. This subgoal can be proved from the gateway requirements.

(2) The gateway stays in the state *connection_ok* also the next k time intervals, $\forall m \leq k.\ ack\ (t + 2 + m) = [connection\_ok]$. This subgoal can be proved using the specification of the gateway architecture. For this purposes an additional lemma about the gateway architecture, *Gateway_L6*, is proved (for details see Section B.7).

(3) At time $t + 3 + k$ the gateway is in the state where the crash data were sent, $ack\ (t+3+k) = [sending\_data]$. This subgoal can be proved from the gateway requirements.

(4) At time $t + 3 + k + d$ the central station receives the crash data (the message list is nonempty), $i\ (t+3+k+d) \neq []$. This subgoal can be proved also from the gateway requirements.

(5) The central station has received no crash data before the time $t + 3 + k + d$ (the message lists are empty), $\forall t2 < (t+3+k+d).\ i\ t2 = []$. This subgoal can be proved using the specification of the gateway architecture. For this purposes an additional lemma about the gateway architecture, *Gateway_L7*, is proved (for details see Section B.7).

(6) The central station has sent no acknowledgment before the time $t + 3 + k + d$, $\forall t3 \leq (t+3+k+d).\ a\ t3 = []$. This subgoal can be proved using the service center specification.

(7) The gateway waits for the acknowledgment from the central station $2d$ time intervals. All this time it stays in the state where the crash data were sent, $\forall x \leq d + d.\ ack\ (t+3+k + x) = [sending\_data])$. This subgoal can be proved using the specification of the gateway architecture. For this purposes an additional lemma about the gateway architecture, *Gateway_L8*, is proved (for details see Section B.7).

Having all these assumptions, we can prove that the gateway establishes the voice communication at time interval $2 * d + t + 4 + k$. The goal is solven using the lemma *GatewaySystem_L1* (see Section B.7).

The lemma *GatewaySystem_L3* says: the voice communication must be established at the time interval $2 * d + t + 4 + k$, if

✓ The predicate *ts* holds for the stream *lose*, and the streams *req*, *a*, *stop* have at every time interval at most one message;

✓ The predicates *Gateway*, *GatewayReq*, and *ServiceCenter* are true for the corresponding streams;

✓ At time t the gateway is in the initial state ($ack\ t = [init\_state]$), and it gets the command to establish the connection with the central station ($req\ (Suc\ t) = [init]$), besides there was no such command before the time t ($\forall t1 < t.\ req\ t1 = []$);

✓ After at least 2 time intervals the gateway get (ones) the command to send the crash data to the central station,
$req\ (t + 3 + k) = [send]$, $\forall m \leq k + 2.\ req\ (t + m) \neq [send]$;

✓ The gateway has received some crash data either at the time interval $t+1$ or at time interval $t + 2$,
$dt\ (Suc\ t) \neq [] \vee dt\ (Suc\ (Suc\ t)) \neq []$;

✓ There is no environment connection problems during the next $2 * d + 4 + k$ time intervals, where d is the communication delay between the gateway and the central station;

lemma *GatewaySystem_L3*:
[*Gateway req dt a stop lose d ack i vc; ServiceCenter i a; msg (Suc 0) req;*
  GatewayReq req dt a stop lose d ack i vc;
  msg (Suc 0) stop; ts lose; msg (Suc 0) a;
  (dt (Suc t) ≠ [] ∨ dt (Suc (Suc t)) ≠ []);
  ack t = [init_state]; req (Suc t) = [init];
  ∀ t1≤t. req t1 = []; ∀ m ≤ k + 2. req (t + m) ≠ [send];
  *req (t + 3 + k) = [send]; ∀ j≤2 * d + (4 + k). lose (t + j) = [False]]*
  ⟹ *vc (2 * d + (t + (4 + k))) = [vc_com]*

The proof of this lemma is analog to the proof of the lemma *GatewaySystem_L2*.

The lemma *Gateway_L6* says: the gateway stays in the state *connection_ok* at the time time intervals $t, \ldots, t + k$, if

✓ The predicate *ts* holds for the stream *lose*, and the streams *req, a, stop* have at every time interval at most one message;

✓ The predicate *Gateway* holds for the corresponding streams;

✓ At time t the gateway is in the state where the connection is established (*ack t = [connection_ok]*, and it gets no command to send the crash data next k time intervals, $\forall m \leq k. \ req \ (t + m) \neq [send]$;

✓ There is no environment connection problems during the next k time intervals;

The lemma *Gateway_L7* says: all time intervals of the output stream i are empty before the time $t + 3 + k + d$, if

✓ The predicate *ts* holds for the stream *lose*, and the streams *req, a, stop* have at every time interval at most one message;

✓ The predicate *Gateway* holds for the corresponding streams;

✓ At time t the gateway is in the initial state, *ack t = [init_state]*, and it gets the command to establish the connection with the central station, *req (Suc t) = [init]*, besides there was no such command before the time t, $\forall t1 \leq t. \ req \ t1 = []$;

✓ After at least 2 time intervals the gateway get (ones) the command to send the crash data to the central station,
req (t + 3 + k) = [send], $\forall m < (k + 3). \ req \ (t + m) \neq [send]$;

✓ There is no environment connection problems during the next $k + d + 3$ time intervals, where d is the communication delay between the gateway and the central station;

The lemma *Gateway_L8* says: the gateway stays in the state "the crash data were sent" next $2d$ time intervals, if

✓ The predicate *ts* holds for the stream *lose*, and the streams *req*, *a*, *stop* have at every time interval at most one message;

✓ The predicate *Gateway* holds for the corresponding streams;

✓ At time t the gateway is in the state where the crash data were sent (*ack* $t = [sending\_data]$);

✓ Until the time $t + d$ no acknowledgment is received, $\forall t3 \leq t + d.\ a\ t3 = []$;

✓ There is no environment connection problems during the next $2d$ time intervals, where d is the communication delay between the gateway and the central station;

4.3.12. Extended Requirements Specification of the Gateway

Proving the lemma *GatewaySystem_L0*, we found out a number of gateway properties which can be seen as requirements to the gateway:

✓ If at the tth point in time the gateway has establish the connection, and it does not get any command to send the E-Call data to the central station until the $(t + k)$th time interval, and also there is no environment connection problems during these time intervals, then it stays it the same state waiting for the command to send the E-Call data (lemma *Gateway_L6*).

✓ If at tth time interval the gateway is in the initial state, and at time interval $t + 1$ the signal about crash comes at first time, and after $3 + m$ time intervals the command to send the crash data comes at first time, and there is no connection fails from the time t until the $(t + 3 + k)$th time interval, then until the $(t + 3 + k + d)$th time interval the output stream i contains no messages (lemma *Gateway_L7*).

✓ If before the tth point in time the gateway has send the E-Call data, but time interval became no acknowledgment from the central station until the $(t + d)$th point in time, and also there is no environment connection problems, then it stays it the same state waiting for the acknowledgment (lemma *Gateway_L8*).

We can add these properties to the specification of the gateway requirements (according to Section 3.1.1). The extended version of the gateway requirements specification *GatewayReqExt* is shown below (the new requirements are marked with green color).

═══ GatewayReqExt(const d ∈ ℕ) ═══════════════════════════════════ timed ═══

in	$req : \{init\_connect\}$; $dt : ECall\_Info$; $a : \{sc\_ack\}$;
	$stop : stop\_vc$; $lose : Bool$
out	$ack : GatewayStatus$; $i : ECall\_Info$; $vc : \{voice\_com\}$

---

univ $\quad k \in \mathbb{N}$

- -

asm $\quad$ $\mathsf{msg}_1(req) \ \wedge \ \mathsf{msg}_1(a) \ \wedge \ \mathsf{msg}_1(stop) \ \wedge \ \mathsf{ts}(lose)$

- -

gar
$\forall\, t \in \mathbb{N}:$

$\quad \mathsf{ti}(ack, t) = \langle init\_state \rangle \ \wedge \ \mathsf{ti}(req, t+1) = \langle init \rangle$
$\quad \wedge \mathsf{ti}(lose, t+1) = \langle \mathsf{false} \rangle \ \wedge \ \mathsf{ti}(lose, t+2) = \langle \mathsf{false} \rangle$
$\quad\quad \rightarrow \mathsf{ti}(ack, t+2) = \langle connection\_ok \rangle$

$\quad \mathsf{ti}(ack, t) = \langle init\_state \rangle \ \wedge \ \mathsf{ti}(req, t+1) = \langle init \rangle \ \wedge \ \mathsf{ti}(req, t+3+k) = \langle send \rangle$
$\quad \wedge \, \forall\, t_1 \leq t : \ \mathsf{ti}(req, t1) = \langle\rangle \ \wedge \ \forall\, m \leq k+3 : \ \mathsf{ti}(req, t+m) \neq \langle send \rangle$
$\quad \wedge \, \forall\, j \leq k+d+3 : \ \mathsf{ti}(lose, t+j) = \langle \mathsf{false} \rangle$
$\quad\quad \dashrightarrow \forall\, t_2 \leq t+3+k+d : \ \mathsf{ti}(i, t_2) = \langle\rangle$

$\quad \mathsf{ti}(ack, t) = \langle connection\_ok \rangle \ \wedge \ \forall\, m \leq k : \ \mathsf{ti}(req, t+m) \neq \langle send \rangle$
$\quad \wedge \, \forall\, j \leq k : \ \mathsf{ti}(lose, t+j) = \langle \mathsf{false} \rangle$
$\quad\quad \dashrightarrow \forall\, y \leq k : \ \mathsf{ti}(ack, t+y) = \langle connection\_ok \rangle$

$\quad \mathsf{ti}(ack, t) = \langle connection\_ok \rangle \ \wedge \ \mathsf{ti}(req, t+1) = \langle send \rangle$
$\quad \wedge (\forall\, k \in \mathbb{N} : \ k \leq d+1 \rightarrow \mathsf{ti}(lose, t+k) = \langle \mathsf{false} \rangle)$
$\quad\quad \rightarrow \mathsf{ti}(i, t+d+1) = \mathsf{last}^{tt}(dt, t) \ \wedge \ \mathsf{ti}(ack, t+1) = \langle sending\_data \rangle$

$\quad \mathsf{ti}(ack, t) = \langle sending\_data \rangle \ \wedge \ \forall\, t_3 \leq t+d : \mathsf{ti}(a, t_3) = \langle\rangle$
$\quad \wedge \, \forall\, j \leq d+d : \ \mathsf{ti}(lose, t+j) = \langle \mathsf{false} \rangle$
$\quad\quad \dashrightarrow \forall\, x \leq d+d : \ \mathsf{ti}(ack, t+x) = \langle sending\_data \rangle$

$\quad \mathsf{ti}(ack, t+d) = \langle sending\_data \rangle \ \wedge \ \mathsf{ti}(a, t+1) = \langle sc\_ack \rangle$
$\quad \wedge \ (\forall\, k \in \mathbb{N} : \ k \leq d+1 \rightarrow \mathsf{ti}(lose, t+k) = \langle \mathsf{false} \rangle)$
$\quad\quad \rightarrow \mathsf{ti}(vc, t+d+1) = \langle vc\_com \rangle$

The specified gateway architecture fulfills certainly the extended requirements, i.e. that the specification *Gateway* is a refinement of the specification by the following *lemma* (the specification *GatewayReqExt* is then translated schematically into the Isabelle/HOL predicate *GatewayReqExt*, see Section B.7):

lemma *Gateway_Ext*:
$\quad$ *Gateway req dt a stop lose d ack i vc*
$\quad \Longrightarrow$
$\quad$ *GatewayReqExt req dt a stop lose d ack i vc*

Using the extended version of the gateway requirements specification, we can do the proofs of system properties *directly from the properties* of the gateway, without extraction the definitions of the gateway architecture and the properties of its components.

The whole Isabelle/HOL proofs of the lemmas *Gateway_Ext*, *GatewaySystem_L0–GatewaySystem_L3*, *Gateway_L6_le*, *Gateway_L7*, *Gateway_L8* and of all other auxiliary lemmas are presented in Section B.7.

4.3.13. Results of the Case Study

In this case study we have shown how we can verify larger systems using the idea of the refinement-based verification. The proofs of a number of system properties were done *directly from the properties* of the gateway, without extraction the definitions of the gateway architecture and the properties of its components. We also present an example of extension of the requirements specification by the new properties (according to Section 3.3).

The FOCUS specifications of all components of the gateway system were translated schematically to Isabelle/HOL and the refinement relation between the requirement and the architecture specification was proved both for the gateway component and for the gateway system. The correctness of the input/output relations was also proved for all components of the system.

4.4. Summary

We have presented in this chapter three case studies that cover different application areas and different specification elements to show feasibility of the approach:

✓ Steam Boiler System (process control),

✓ FlexRay communication protocol (data transmission),

✓ Automotive-Gateway System (memory and processing components, data transmission).

The following has been done within every case study:

✓ The FOCUS specifications of all components of the system have been translated schematically to Isabelle/HOL and the refinement relation between the requirement and the architecture specification of the system has been proved.

✓ The correctness of the input/output relations has been also proved for all components of the system (automatically, according to the proof schemata from Sections 2.11.4 and 2.13.3).

Proving the refinement relation for the steam boiler system specifications in Isabelle/HOL, we found out that to argue about properties of the Controller component of the system we need an additional (wrt. the original specification from [BS01]) assumption about the input stream y – that the stream y is a time-synchronous one.

The proofs for the steam boiler system take ca. 200 lop[9], the proofs for the FlexRay communication protocol take ca. 300 lop – for both these systems no extra decomposition of proofs and components is needed. The proofs for the gateway system are more complicated and take ca. 1700 lop – using two decomposition layers (the architecture specification of the whole system and the architecture specification of the gateway component) instead a single one we have got a clear proof structure, applying the ideas of refinement layers (see Section 3.1.1) we can also reuse the proofs about the gateway component later.

In these case studies we have shown

✓ how to reformulate FOCUS constructions which are not very well situated to the direct translation to Isabelle/HOL and without changing their semantics;

✓ how we can deal with local variables (states);

✓ in which way we can represent mutual recursive functions to avoid problems in proofs;

✓ how we can deal with sheaves of channels ans parameters, as well as with specification replications;

✓ how we can verify larger systems using the idea of the refinement-based verification.

---

[9]lines of proof

5. Conclusions

This chapter presents the summary of the key contributions of the thesis and gives a short overview of future work.

The increasing complexity and the safety and quality requirements of embedded real-time systems implies that it is insufficient only to test them, because testing can only demonstrate the presence, but not the absence of errors. Formal methods can provide the level of assurance required by the increasing complexity and the high safety and quality requirements to such systems, because these allow not only to test correctness and safety, but also to prove them: verification guarantees fulfillment of the requirements. A formal specification is more precise than a natural language one, but it can also contain mistakes or disagree with requirements. Therefore, for safety critical systems it is not enough to have detached formal specifications – in this case formal verification is needed. This is the only way to be sure that the specification conforms to its requirements and is consistent.

In this thesis we have introduced the coupling of the formal specification framework FOCUS in the generic theorem prover Isabelle/HOL with focus on specification and verification of systems that are especially safety critical – embedded real-time systems.

5.1. Summary

The result of the coupling of the formal specification framework FOCUS in the generic theorem prover Isabelle/HOL is the framework "FOCUS on Isabelle". By considering this framework we can influence the complexity of proofs and their reusability already during the specification phase, because the specification and verification/validation methodologies are treated here as a single, joined, methodology with the main focus on the specification part. Given a system, represented in a formal specification framework, we can verify its properties by translating the specification to a Higher-Order Logic and subsequently using the theorem prover Isabelle/HOL (or the point of disagreement will be found).

The key contributions of the thesis are

✓ Deep embedding of that part of the framework FOCUS, which is appropriate for specification of real-time systems, into Isabelle/HOL:

  ◇ representation of FOCUS datatypes in Isabelle/HOL,

  ◇ representation of FOCUS streams in Isabelle/HOL,

  ◇ representation of the FOCUS operators on streams:

    * length of a stream,

                * nth message of a stream,
                * concatenation operator,
                * prefix of a stream,
                * truncation operator,
                * domain and range of a stream,
                * time stamp operator, and
                * stuttering removal operator.
            ◇ specification semantics and techniques,
            ◇ representation of the FOCUS extras:
                * encapsulated states (local variables, control states, oracles),
                * sheaf of channels and specification replication.

✓ Syntax extensions for FOCUS for the argumentation over time intervals:

            ◇ a special kind of FOCUS tables – tiTable,
            ◇ timed state transition diagrams (TSTDs),
            ◇ a number of new operators, such as:
                * time interval operator,
                * timed merge,
                * timed truncation operator,
                * limited number of messages per time interval,
                * stuttering removal operator for timed streams,
                * changing time granularity,
                * deleting the first time interval,
                * the last nonempty time interval until some time interval, and
                * number of time intervals in a finite timed stream.

The deep embedding into Isabelle/HOL includes all these extensions.

✓ A number of Isabelle/HOL theories and the corresponding schemata to prove correctness of the relations between the sets of input, output and local channels of a specified system. These proof schemata for the correctness properties are standard and can be used automatically. If the proof fails, the specification of the corresponding set is incorrect and must be changed.

✓ The specification and verification/validation methodology, which enables to validate and verify the system specifications in a methodological way. Using this approach we can validate the refinement relation between two given systems. This methodology uses particularly the idea of refinement-based verification, where a verification of system properties can be treated as a validation of a system specification with respect to the specification representing the properties. Thus, designing a system, a refinement relation must be shown not only between requirements and architecture specifications, but for every step at which a more abstract specification is refined to a more precise one. The proofs of the refinement relations between specifications of neighbor levels of abstractions are in general simpler and shorter as the proof of the refinement relation between the more abstract and the more concrete specifications.

The feasibility of this approach was evaluated on three case studies that cover different application areas:

✓ Steam Boiler System (process control),

✓ FlexRay communication protocol (data transmission),

✓ Automotive-Gateway System (memory and processing components, data transmission)

Scalability of the investment of time needed for the verification of a concrete property of a concrete system is another interesting point. As mentioned above, the size and reusability of proofs depend on the nature of specifications and on the granularity of the refinement steps – by proving the properties of subcomponents first, we can reuse the results later. Starting from the proof of properties of some large system, we also will find out the needed subcomponent properties. We add them to the requirements specifications of the subcomponent properties, in order to have possibility to reuse them later easily (see Section 3.3, 4.3.12). Thus, the modularity and reusability play here a major role.

Isabelle/HOL is an interactive semi-automatic theorem prover and the time needed for the verification of system properties is also human depended. To simplify the proof process the clear overview of the (sub)components and their properties is needed – this must be established during the specification phase. It is sufficient and worth while to spend more time into the specification phase, because this will reduce the time needed for the verification extensively.

The results of "Focus on Isabelle" can be also extended to a complementary approach, "Janus on Isabelle", that represents a coupling of a Janus with Isabelle/HOL. Janus is a specification framework for services, which is developed on the base of Focus and uses different, but similar syntax and semantics.

5.2. Outlook

We have presented here how one can influence the complexity of proofs and their reusability already doing the specification in Focus – avoid the untimed specifications, specify the behavior via time intervals to express the causality property (weak or strong) explicitly, use extended operator definitions, etc. However, the results of our approach give rise to another interesting research challenges:

✓ Can we define on the basis of the presented specification syntax some proof schemata or, more concrete, Isabelle/HOL tactics which can be applied automatically?

✓ Is it possible to have such a schema for every syntax construction?

✓ How can we find out the optimal decomposition of a component into subcomponents in order to get simpler proofs?

Another promising topic which was only touched in this thesis is the formal refinement of timing properties, e.g. changing time granularity. We have presented here in which way this kind of operators on timed streams can be defined as well as how the proofs of their properties can be done. Based on this results the formal theory of time refinement can be build.

An interesting research area where the findings of the thesis also can be extended is an enhancement of the specification technics and the refinement relations between them. The first step in this direction is the application of the results of the thesis, especially the syntax extensions for FOCUS, to elaborate a model-based process supporting structured development for the CoCoME Modelling Contest (Common Component Modelling Example, see [Auta] and [BFH+07]).

Bibliography

[Abr96]   J.-R. Abrial. *The B-book: assigning programs to meanings.* Cambridge University Press, New York, NY, USA, 1996.

[Age94]   S. Agerholm. *A HOL Basis for Reasoning about Functional Programs.* PhD thesis, University of Aarhus, 1994.

[Auta]   Modelling Contest: Common Component Modelling Example (Co-CoME). http://agrausch.informatik.uni-kl.de/CoCoME/.

[Autb]   Verisoft–Automotive Project. http://www4.in.tum.de/~verisoft/automotive.

[BDD+92]   Manfred Broy, Frank Dederich, Claus Dendorfer, Max Fuchs, Thomas Gritzner, and Rainer Weber. The Design of Distributed Systems – An Introduction to FOCUS. Technical Report TUM-I9202, Technische Univerität München, 1992.

[Ber05]   S. Berghofer. Isabelle/HOL Filter Theory, 2005.

[BFH+07]   Manfred Broy, Jorge Fox, Florian Hölzl, Dagmar Koss, Michael Meisinger Marco Kuhrmann, Birgit Penzenstadler, Sabine Rittmann, Bernhard Schätz, Maria Spichkova, and Doris Wild. *Modeling CoCoME with Focus/AutoFocus.* LNCS. Springer, 2007.

[Bro97]   M. Broy. Compositional Refinement of Interactive Systems. *J. ACM*, 44(6):850–891, 1997.

[Bro01]   Manfred Broy. Refinement of time. *Theor. Comput. Sci.*, 253(1):3–26, 2001.

[Bro04]   Manfred Broy. Time, Abstraction, Causality and Modularity in Interactive Systems: Extended Abstract. *Electr. Notes Theor. Comput. Sci.*, 108:3–9, 2004.

[Bro05]   Manfred Broy. Service-oriented Systems Engineering: Specification and design of services and layered architectures. The JANUS Approach. pages 47–81, July 2005.

[BS01]   M. Broy and K. Stølen. *Specification and Development of Interactive Systems: Focus on Streams, Interfaces, and Refinement.* 2001.

[CP96]   Ching-Tsun Chou and D. Peled. Formal Verification of a Partial-Order Reduction Technique for Model Checking. In *TACAS*, pages 241–257, 1996.

[DGM97]   M. Devillers, D. Griffioen, and O. Müller. Possibly Infinite Sequences in Theorem Provers: A Comparative Study. In Elsa Gunther, editor, *Theorem Proving in Higher Order Logics (TPHOL'97)*, number LNCS 1275. Springer-Verlag, 1997.

[Fle]   FlexRay Consortium. http://www.flexray.com.

[Fle04]   FlexRay Consortium. *FlexRay Communication System - Protocol Specification - Version 2.0*, 2004.

[GKRB96]   Radu Grosu, Cornel Klein, Bernhard Rumpe, and Manfred Broy. State Transition Diagrams. Technical Report TUM-I9630, Technische Univerität München, 1996.

[GR06]   B. Gajanovic and B. Rumpe. Isabelle/HOL-Umsetzung strombasierter Definitionen zur Verifikation von verteilten, asynchron kommunizierenden Systemen. Technical Report Informatik-Bericht 2006-03, Technische Universität Braunschweig, 2006.

[HMP91]   Thomas A. Henzinger, Zohar Manna, and Amir Pnueli. Timed Transition Systems. In *REX Workshop*, pages 226–251, 1991.

[JR97]   B. Jacobs and J. Rutten. A Tutorial on (Co)Algebras and (Co)Induction. *Bulletin of the EATCS*, 62:222–259, 1997.

[KPR97]   C. Klein, C. Prehofer, and B. Rumpe. Feature Specification and Refinement with State Transition Diagrams. In P. Dini, editor, Fourth IEEE Workshop on Feature Interactions in Telecommunications Networks and Distributed Systems. IOS-Press, 1997., 1997.

[KS06a]   C. Kühnel and M. Spichkova. FlexRay und FTCom: Formale Spezifikation in FOCUS. Technical Report TUM-I0601, Technische Universität München, 2006.

[KS06b]   C. Kühnel and M. Spichkova. Upcoming Automotive Standards for Fault-Tolerant Communication: FlexRay and OSEKtime FT-Com. In *EFTS 2006 International Workshop on Engineering of Fault Tolerant Systems*. Universite du Luxembourg, CSC: Computer Science and Communication, 2006.

[KSed]   C. Kühnel and M. Spichkova. Fault-Tolerant Communication for Distributed Embedded Systems. In *Software Engineering and Fault Tolerance*, Series on Software Engineering and Knowledge Engineering, 2007 (to be published).

[LSBB92]   R. Letz, J. Schumann, S. Bayerl, and W. Bibel. SETHEO: a high-performance theorem prover. *J. Autom. Reason.*, 8(2):183–212, 1992.

[MNvOS99] O. Müller, T. Nipkow, D. von Oheimb, and O. Slotosch. HOLCF = HOL + LCF. *Journal of Functional Programming*, (9(2)):191–223, 1999.

[Nip05] T. Nipkow. Theory List. http://isabelle.in.tum.de/library/HOL/List.html, 2005.

[NPW02] T. Nipkow, L. C. Paulson, and M. Wenzel. *Isabelle/HOL – A Proof Assistant for Higher-Order Logic*, volume 2283 of *LNCS*. Springer, 2002.

[NS95] T. Nipkow and K. Slind. I/O Automata in Isabelle/HOL. In P. Dybjer, editor, *Proc. of Types for Proofs and Programs*, number LNCS 996, 1995.

[Pau94] L. C. Paulson. *Isabelle: A Generic Theorem Prover*, volume 828 of *LNCS*. Springer, 1994.

[Reg94] F. Regensburger. *HOLCF: eine konservative Erweiterung von HOL um LCF*. PhD thesis, Technische Universität München, 1994.

[Reg95] F. Regensburger. HOLCF: Higher Order Logic of Computable Functions. In *TPHOLs*, pages 293–307, 1995.

[Rob91] Robert Bosch GmbH. *CAN Specification Version 2.0*, 1991.

[SB99] J. Schumann and M. Breitling. Formalisierung und Beweis einer Verfeinerung aus FOCUS mit automatischen Theorembeweisern – Fallstudie. Technical Report TUM-I9904, Technische Universität München, 1999.

[Slo97] Oscar Slotosch. *Refinements in HOLCF: Implementation of Interactive Systems*. PhD thesis, Technische Universität München, 1997.

[Spi03] M. Spichkova. A coalgebraic view at data flow systems. Master's thesis, Technische Universität Dresden, 2003.

[Spi06] M. Spichkova. FlexRay: Verification of the FOCUS Specification in Isabelle/HOL. A Case Study. Technical Report TUM-I0602, Technische Universität München, 2006.

[SS95] Bernhard Schätz and Katharina Spies. Formale Syntax zur logischen Kernsprache der FOCUS-Entwicklungsmethodik. Technical Report TUM-I9529, Technische Universität München, 1995.

[Ver] Verisoft Project. http://www.verisoft.de.

[vO] D. von Oheimb. Theory Nat_Infinity. http://isabelle.in.tum.de/library/HOL/Library/Nat_Infinity.html.

[vO05]   D. von Oheimb.   Isabelle/HOLCF Formalization of FOCUS Streams. `http://isabelle.in.tum.de/library/HOLCF/FOCUS/` `Fstream.html`, 2005.

[Wen04]   M. Wenzel. *The Isabelle/Isar Reference Manual.* Technische Universität München, 2004. Part of the Isabelle distribution.

[Zha06]   Bo Zhang.   On the Formal Verification of the FlexRay Communication Protocol. In *Automatic Verification of Critical Systems (AVoCS)*, pages 184–189, 2006.

Appendix A.

Isabelle Definitions and Lemmas about FOCUS Operators

A.1. Theory stream.thy (FOCUS streams)

theory *stream = Main + ListExtras + ListLemmas + ArithExtras + Filter:*

types *'a fstream = 'a list list*

types *'a istream = nat ⇒ 'a list*

types *'a iustream = nat ⇒ 'a*

datatype *'a stream = FinT 'a fstream*
   | FinU 'a list
   | InfT 'a istream
   | InfU 'a iustream

consts
   nticks :: nat ⇒ 'a fstream
   finU_dom :: 'a list ⇒ nat set
   finT_range :: 'a fstream ⇒ 'a set
   fin_find1nonemp :: 'a fstream ⇒ 'a list
   fin_find1nonemp_index :: 'a fstream ⇒ nat
   fin_length :: 'a fstream ⇒ nat
   fin_nth :: 'a fstream ⇒ nat ⇒ 'a
   inf_nth :: 'a istream ⇒ nat ⇒ 'a
   inf_prefix :: 'a list ⇒ (nat ⇒ 'a) ⇒ nat ⇒ bool
   fin_truncate :: 'a list ⇒ nat ⇒ 'a list
   inf_truncate :: (nat ⇒ 'a) ⇒ nat ⇒ 'a list
   fin_msg :: nat ⇒ 'a list list ⇒ bool
   inf_make_untimed1 :: 'a istream ⇒ nat ⇒ 'a
   fin_tm :: 'a fstream ⇒ nat ⇒ nat
   inf_tm :: ('a istream × nat) ⇒ nat
   fst_remdups :: 'a list ⇒ 'a list
   inf_inf_remdups :: (nat ⇒ 'a) ⇒ (nat ⇒'a)
   fin_get_prefix :: ('a fstream × nat) ⇒ 'a fstream
   infT_get_prefix :: ('a istream × nat) ⇒ 'a fstream
   infU_get_prefix :: (nat ⇒ 'a) ⇒ nat ⇒ 'a list
   fin_merge_ti :: 'a fstream ⇒ 'a fstream ⇒ 'a fstream
   fin_last_ti :: ('a list) list ⇒ nat ⇒ 'a list
   inf_last_ti :: 'a istream ⇒ nat ⇒ 'a list

constdefs
   finU_dom_inat :: 'a list ⇒ inat set

$finU\_dom\_inat\ s \equiv \{x.\ \exists\ i.\ x = (Fin\ i) \land i < (length\ s)\}$

constdefs
  $infU\_dom :: inat\ set$
  $infU\_dom \equiv \{x.\ \exists\ i.\ x = (Fin\ i)\} \cup \{\infty\}$

primrec
  $nticks\ 0 = []$
  $nticks\ (Suc\ i) = [] \mathbin{\#} (nticks\ i)$

primrec
  $finU\_dom\ [] = \{\}$
  $finU\_dom\ (x\mathbin{\#}xs) = \{length\ xs\} \cup (finU\_dom\ xs)$

primrec
  $finT\_range\ [] = \{\}$
  $finT\_range\ (x\mathbin{\#}xs) = (set\ x) \cup finT\_range\ xs$

constdefs
  $finU\_range :: {}'a\ list \Rightarrow {}'a\ set$
  $finU\_range\ x \equiv set\ x$

constdefs
  $infT\_range :: {}'a\ istream \Rightarrow {}'a\ set$
  $infT\_range\ s \equiv \{y.\ \exists\ i::nat.\ y\ mem\ (s\ i)\}$

constdefs
  $infU\_range :: (nat \Rightarrow {}'a) \Rightarrow {}'a\ set$
  $infU\_range\ s \equiv \{\ y.\ \exists\ i::nat.\ y = (s\ i)\ \}$

constdefs
  $stream\_range :: {}'a\ stream \Rightarrow {}'a\ set$
  $stream\_range\ s \equiv case\ s\ of$
    $FinT\ x \Rightarrow finT\_range\ x$
    $\mid\ FinU\ x \Rightarrow finU\_range\ x$
    $\mid\ InfT\ x \Rightarrow infT\_range\ x$
    $\mid\ InfU\ x \Rightarrow infU\_range\ x$

constdefs
  $inf\_tl :: (nat \Rightarrow {}'a) \Rightarrow (nat \Rightarrow {}'a)$
  $inf\_tl\ s \equiv (\lambda\ i.\ s\ (Suc\ i))$

primrec
  $fin\_find1nonemp\ [] = []$
  $fin\_find1nonemp\ (x\mathbin{\#}xs) =$
    $(\ if\ x = []$
      $then\ fin\_find1nonemp\ xs$
      $else\ x\ )$

constdefs
  $inf\_find1nonemp :: {}'a\ istream \Rightarrow {}'a\ list$
  $inf\_find1nonemp\ s$
    $\equiv$
  $(\ if\ (\exists\ i.\ s\ i \neq [])$
    $then\ s\ (LEAST\ i.\ s\ i \neq [])$

   else [])

primrec
 fin_find1nonemp_index [] *= 0*
 fin_find1nonemp_index (*x#xs*) *=*
  (*if x =* []
   then Suc (*fin_find1nonemp_index xs*)
   else 0)

constdefs
 inf_find1nonemp_index :: *'a istream* ⇒ *nat*
 inf_find1nonemp_index s
 ≡
 (*if* (∃ *i. s i ≠* [])
  then (*LEAST i. s i ≠* [])
  else 0)

constdefs
 inf_drop :: *nat* ⇒ (*nat* ⇒ *'a*) ⇒ (*nat* ⇒ *'a*)
 inf_drop i s ≡ λ *j. s* (*i+j*)

primrec
 fin_length [] *= 0*
 fin_length (*x#xs*) *=* (*length x*) *+* (*fin_length xs*)

constdefs
 stream_length :: *'a stream* ⇒ *inat*
 stream_length s ≡
  case s of (*FinT x*) ⇒ *Fin* (*fin_length x*)
   | (*FinU x*) ⇒ *Fin* (*length x*)
   | (*InfT x*) ⇒ ∞
   | (*InfU x*) ⇒ ∞

primrec
 fin_nth_Cons:
 fin_nth (*hds # tls*) *k =*
  (*if hds =* []
   then fin_nth tls k
   else (*if* (*k < (length hds*))
    then nth hds k
    else fin_nth tls (*k − length hds*)))

primrec
 inf_nth s 0 =
  hd (*s* (*LEAST i.(s i) ≠* []))

 inf_nth s (*Suc k*) *=*
  (*if* ((*Suc k*) *< (length (s 0*)))
   then (*nth* (*s 0*) (*Suc k*))
   else (*if* (*s 0*) *=* []
    then (*inf_nth* (*inf_tl* (*inf_drop*
     (*LEAST i.* (*s i*) *≠* []) *s*)) *k*)
    else inf_nth (*inf_tl s*) *k*))

constdefs

$\quad$ *stream_nth* :: *'a stream* $\Rightarrow$ *nat* $\Rightarrow$ *'a*

$\quad$ *stream_nth s k* $\equiv$

$\qquad$ *case s of (FinT x)* $\Rightarrow$ *fin_nth x k*

$\qquad\quad$ | *(FinU x)* $\Rightarrow$ *nth x k*

$\qquad\quad$ | *(InfT x)* $\Rightarrow$ *inf_nth x k*

$\qquad\quad$ | *(InfU x)* $\Rightarrow$ *x k*

primrec

$\quad$ *inf_prefix* [] *s k = True*

$\quad$ *inf_prefix (x#xs) s k =*

$\qquad$ (*((x = (s k)) $\wedge$ (inf_prefix xs s (Suc k)))*

constdefs

$\quad$ *stream_prefix* :: *'a stream* $\Rightarrow$ *'a stream* $\Rightarrow$ *bool*

$\quad$ *stream_prefix p s* $\equiv$

$\quad$ *(case p of*

$\qquad$ *(FinT x)* $\Rightarrow$

$\qquad$ *(case s of (FinT y)* $\Rightarrow$ *(x $\leq$ y)*

$\qquad\qquad$ | *(FinU y)* $\Rightarrow$ *False*

$\qquad\qquad$ | *(InfT y)* $\Rightarrow$ *inf_prefix x y 0*

$\qquad\qquad$ | *(InfU y)* $\Rightarrow$ *False*)

$\qquad$ | *(FinU x)* $\Rightarrow$

$\qquad$ *(case s of (FinT y)* $\Rightarrow$ *False*

$\qquad\qquad$ | *(FinU y)* $\Rightarrow$ *(x $\leq$ y)*

$\qquad\qquad$ | *(InfT y)* $\Rightarrow$ *False*

$\qquad\qquad$ | *(InfU y)* $\Rightarrow$ *inf_prefix x y 0*)

$\qquad$ | *(InfT x)* $\Rightarrow$

$\qquad$ *(case s of (FinT y)* $\Rightarrow$ *False*

$\qquad\qquad$ | *(FinU y)* $\Rightarrow$ *False*

$\qquad\qquad$ | *(InfT y)* $\Rightarrow$ *($\forall$ i. x i = y i)*

$\qquad\qquad$ | *(InfU y)* $\Rightarrow$ *False*)

$\qquad$ | *(InfU x)* $\Rightarrow$

$\qquad$ *(case s of (FinT y)* $\Rightarrow$ *False*

$\qquad\qquad$ | *(FinU y)* $\Rightarrow$ *False*

$\qquad\qquad$ | *(InfT y)* $\Rightarrow$ *False*

$\qquad\qquad$ | *(InfU y)* $\Rightarrow$ *($\forall$ i. x i = y i)*))

primrec

$\quad$ *fin_truncate* [] *n =* []

$\quad$ *fin_truncate (x#xs) i =*

$\qquad$ *(case i of 0* $\Rightarrow$ []

$\qquad\quad$ | *(Suc n)* $\Rightarrow$ *x # (fin_truncate xs n))*

constdefs

$\quad$ *fin_truncate_plus* :: *'a list* $\Rightarrow$ *inat* $\Rightarrow$ *'a list*

$\quad$ *fin_truncate_plus s n*

$\quad$ $\equiv$

$\quad$ *case n of (Fin i)* $\Rightarrow$ *fin_truncate s i*

$\qquad\quad$ | ∞ $\quad$ $\Rightarrow$ *s*

primrec

$\quad$ *inf_truncate s 0 =* [*s 0*]

$\quad$ *inf_truncate s (Suc k) = (inf_truncate s k)* @ [*s (Suc k)*]

constdefs

inf_truncate_plus :: *'a istream* ⇒ *inat* ⇒ *'a stream*
inf_truncate_plus s n
≡
case n of (Fin i) ⇒ *FinT (inf_truncate s i)*
         | ∞   ⇒ InfT s

constdefs
   fin_inf_append ::
      'a list ⇒ *(nat* ⇒ *'a)* ⇒ *(nat* ⇒ *'a)*
   fin_inf_append us s ≡
      (λ i. (if (i < (length us))
      then (nth us i)
      else s (i − (length us))))

constdefs
   ts :: *'a istream* ⇒ *bool*
   ts s ≡ ∀ i. (length (s i) = 1)

constdefs
   msg :: *nat* ⇒ *'a istream* ⇒ *bool*
   msg n s ≡ ∀ t. length (s t) ≤ n

primrec
fin_msg n [] = True
fin_msg n (x#xs) = (((length x) ≤ n) ∧ (fin_msg n xs))

constdefs
   fin_make_untimed :: *'a fstream* ⇒ *'a list*
   fin_make_untimed x ≡ concat x

primrec
inf_make_untimed1_0:
   inf_make_untimed1 s 0 =
      hd (s (LEAST i.(s i) ≠ []))

inf_make_untimed1_Suc:
   inf_make_untimed1 s (Suc k) =
   (if ((Suc k) < length (s 0))
      then nth (s 0) (Suc k)
      else (if (s 0) = []
            then (inf_make_untimed1 (inf_tl (inf_drop
                  (LEAST i. ∀ j. j < i ⟶ (s j) = [])
                  s)) k)
            else inf_make_untimed1 (inf_tl s) k))

constdefs
   inf_make_untimed :: *'a istream* ⇒ *(nat* ⇒ *'a)*
   inf_make_untimed s
   ≡
   λ i. inf_make_untimed1 s i

constdefs
 make_untimed :: *'a stream* ⇒ *'a stream*
 make_untimed s ≡
  case s of (*FinT x*) ⇒ *FinU* (*fin_make_untimed x*)
   | (*FinU x*) ⇒ *FinU x*
   | (*InfT x*) ⇒
    (*if* (∃ *i*.∀ *j*. *i* < *j* ⟶ (*x j*) = [])
     then FinU (*fin_make_untimed* (*inf_truncate x*
       (*LEAST i*.∀ *j*. *i* < *j* ⟶ (*x j*) = []))))
     else InfU (*inf_make_untimed x*))
   | (*InfU x*) ⇒ *InfU x*

primrec
 fin_tm [] *k* = *k*
 fin_tm (*x#xs*) *k* =
  (*if k* = *0*
  then 0
  else (*if* (*k* ≤ *length x*)
    then (*Suc 0*)
    else Suc(*fin_tm xs* (*k* − *length x*))))

lemma *inf_tm_hint1*:
⟦ *i2* = *Suc i* − *length a*; ¬ *Suc i* ≤ *length a*; *a* ≠ [] ⟧
⟹ *i2* < *Suc i*
 by *auto*

lemma *inf_tm_hint*:
⟦ *i2* = *Suc i* − *length* ((*s*::*'a istream*) (*LEAST x WRT* (λ*n*. *n*). *s x* ≠ []));
 ¬ *Suc i* ≤ *length* (*s* (*LEAST x WRT* (λ*n*. *n*). *s x* ≠ [])); ∃*j*. *s j* ≠ [] ⟧
⟹ *i2* < *Suc i*
 apply (*rule inf_tm_hint1*, *assumption*+)
 apply (*erule exE*, *rule LeastM_natI*, *assumption*)
 done

recdef *inf_tm measure*(λ(*s,n*). *n*)
 inf_tm (*s*, *0*) = *0*

 inf_tm (*s*, *Suc i*) =
 (*if* (∀ *j*. *s j* = [])
  then 0
  else
   (*let*
    k = (*LEAST x WRT* (λ*n*. *n*). *s x* ≠ [])
   in
    (*if* (*Suc i*) ≤ (*length*(*s k*))
    then (*Suc k*)
    else (*let*
      i2 = (*Suc i*) − (*length* (*s k*));
      s2 = *inf_drop* (*Suc k*) *s*
     in
      inf_tm (*s2*, *i2*))
))
)
(**hints** *intro*: *inf_tm_hint* [*rule_format*])

constdefs

  finT_filter :: *'a set* $\Rightarrow$ *'a fstream* $\Rightarrow$ *'a fstream*
  finT_filter m s $\equiv$ *map* (λ *s. filter* (λ *y. y* $\in$ *m*) *s*) *s*

  infT_filter :: *'a set* $\Rightarrow$ *'a istream* $\Rightarrow$ *'a istream*
  infT_filter m s $\equiv$ (λi.(*filter* (λ *x. x* $\in$ *m*) (*s i*)))

constdefs

  finT_remdups :: *'a fstream* $\Rightarrow$ *'a fstream*
  finT_remdups s $\equiv$ *map* (λ *s. remdups s*) *s*

  infT_remdups :: *'a istream* $\Rightarrow$ *'a istream*
  infT_remdups s $\equiv$ (λi.(*remdups* (*s i*)))

primrec

fst_remdups [] = []
fst_remdups (*x#xs*) =
  (*if xs* = []
  then [*x*]
  else (*if x* = (*hd xs*)
     then fst_remdups xs
     else (*x#xs*)))

constdefs

  fst_inf_inf_remdups :: (*nat* $\Rightarrow$ *'a*) $\Rightarrow$ (*nat* $\Rightarrow$ *'a*)
  fst_inf_inf_remdups s
  $\equiv$ *drop_seq* (*LEAST i.* $\forall$ *j. j* < *i* $\longrightarrow$ *s i* = *s j*) *s*

constdefs

  fst_inf_remdups :: (*nat* $\Rightarrow$ *'a*) $\Rightarrow$ *'a stream*
  fst_inf_remdups s
  $\equiv$
  (*if* ($\forall$ *i.* (*s 0*) = (*s i*))
  then FinU [*s 0*]
  else InfU (*fst_inf_inf_remdups s*))

primrec

inf_inf_remdups s 0 = *s 0*
inf_inf_remdups s (*Suc i*) =
  (*let x* = (*fst_inf_inf_remdups s*)
  in
  inf_inf_remdups (*drop_seq* (*Suc 0*) *x*) *i*)

constdefs

  inf_remdups :: (*nat* $\Rightarrow$ *'a*) $\Rightarrow$ *'a stream*
  inf_remdups s $\equiv$
  (*if* ($\exists$ *i.* $\forall$ *j. i* $\leq$ *j* $\longrightarrow$ (*s i*) = (*s j*))
  then FinU (*remdups* (*inf_truncate s*
     (*LEAST i.* $\forall$ *j. i* $\leq$ *j* $\longrightarrow$ (*s i*) = (*s j*))))
  else InfU (*inf_inf_remdups s*))

constdefs

```
ti :: 'a fstream ⇒ nat ⇒ 'a list
ti s i ≡
  (if s = []
   then []
   else (nth s i))
```

constdefs
```
CorrectSheaf :: nat ⇒ bool
CorrectSheaf n ≡ 0 < n
```

constdefs
```
inf_disjS :: 'b set ⇒ ('b ⇒ 'a istream) ⇒ bool
inf_disjS IdSet nS
  ≡
∀ (t::nat) i j. (i:IdSet) ∧ (j:IdSet) ∧
((nS i) t) ≠ [] ⟶ ((nS j) t) = []
```

constdefs
```
inf_disj :: nat ⇒ (nat ⇒ 'a istream) ⇒ bool
inf_disj n nS
  ≡
∀ (t::nat) (i::nat) (j::nat).
i < n ∧ j < n ∧ i ≠ j ∧ ((nS i) t) ≠ [] ⟶
((nS j) t) = []
```

recdef
```
fin_get_prefix measure(λ(s,n). length s + n)
fin_get_prefix([], n) = []
fin_get_prefix(x#xs, i) =
  ( if (length x) < i
    then x # fin_get_prefix(xs, (i − (length x)))
    else [take i x] )
```

constdefs
```
fin_get_prefix_plus :: 'a fstream ⇒ inat ⇒ 'a fstream
fin_get_prefix_plus s n
  ≡
case n of (Fin i) ⇒ fin_get_prefix(s, i)
        | ∞      ⇒ s
```

lemma *length_inf_drop_hint1*:
```
s k ≠ [] ⟹ length (inf_drop k s 0) ≠ 0
  by (simp add: inf_drop_def)
```

lemma *length_inf_drop_hint2*:
```
(s 0 ≠ [] ⟶ length (inf_drop 0 s 0) < Suc i
  ⟶ Suc i − length (inf_drop 0 s 0) < Suc i)
  by (simp add: inf_drop_def list_length_hint1)
```

recdef
```
infT_get_prefix measure(λ(s,n). n)
```

$infT\_get\_prefix(s,\ 0) = []$

$infT\_get\_prefix(s,\ Suc\ i) =$
$(\ if\ (s\ 0) = []$
$\quad then\ (\ if\ (\forall\ i.\ s\ i = [])$
$\qquad\quad then\ []$
$\qquad\quad else\ (let$
$\qquad\qquad\qquad k = (LEAST\ k.\ s\ k \neq [] \wedge (\forall\ i.\ i < k \longrightarrow s\ i = []));$
$\qquad\qquad\qquad s2 = inf\_drop\ (k{+}1)\ s$
$\qquad\qquad\quad in\ (if\ (length\ (s\ k){=}0)$
$\qquad\qquad\qquad\quad then\ []$
$\qquad\qquad\qquad\quad else\ (if\ (length\ (s\ k) < (Suc\ i))$
$\qquad\qquad\qquad\qquad\quad then\ s\ k\ \#\ infT\_get\_prefix\ (s2,Suc\ i\ -\ length\ (s\ k))$
$\qquad\qquad\qquad\qquad\quad else\ [take\ (Suc\ i)\ (s\ k)]\ )))$
$\qquad\ )$
$\quad else$
$\quad (if\ ((length\ (s\ 0)) < (Suc\ i))$
$\quad\ then\ (s\ 0)\ \#\ infT\_get\_prefix(\ inf\_drop\ 1\ s,\ (Suc\ i)\ -\ (length\ (s\ 0)))$
$\quad\ else\ [take\ (Suc\ i)\ (s\ 0)]$
$\quad\ )$
$)$
(**hints** *recdef_simp: list_length_hint1 Add_Less*)

primrec
$infU\_get\_prefix\ s\ 0 = []$
$infU\_get\_prefix\ s\ (Suc\ i)$
$\quad = (infU\_get\_prefix\ s\ i)\ @\ [s\ i]$

constdefs
$infT\_get\_prefix\_plus :: {}'a\ istream \Rightarrow inat \Rightarrow {}'a\ stream$
$infT\_get\_prefix\_plus\ s\ n$
$\equiv$
$case\ n\ of\ (Fin\ i) \Rightarrow FinT\ (infT\_get\_prefix(s,\ i))$
$\qquad\quad |\ \infty \quad \Rightarrow InfT\ s$

constdefs
$infU\_get\_prefix\_plus :: (nat \Rightarrow {}'a) \Rightarrow inat \Rightarrow {}'a\ stream$
$infU\_get\_prefix\_plus\ s\ n$
$\equiv$
$case\ n\ of\ (Fin\ i) \Rightarrow FinU\ (infU\_get\_prefix\ s\ i)$
$\qquad\quad |\ \infty \quad \Rightarrow InfU\ s$

constdefs
$take\_plus :: inat \Rightarrow {}'a\ list \Rightarrow {}'a\ list$
$take\_plus\ n\ s$
$\equiv$
$case\ n\ of\ (Fin\ i) \Rightarrow (take\ i\ s)$
$\qquad\quad |\ \infty \quad \Rightarrow s$

constdefs
$get\_prefix :: {}'a\ stream \Rightarrow inat \Rightarrow {}'a\ stream$
$get\_prefix\ s\ k \equiv$
$\quad case\ s\ of\ (FinT\ x) \Rightarrow FinT\ (fin\_get\_prefix\_plus\ x\ k)$
$\qquad\qquad |\ (FinU\ x) \Rightarrow FinU\ (take\_plus\ k\ x)$
$\qquad\qquad |\ (InfT\ x) \Rightarrow infT\_get\_prefix\_plus\ x\ k$

$$| \ (InfU \ x) \Rightarrow infU\_get\_prefix\_plus \ x \ k$$

primrec
fin_merge_ti [] *y* = *y*
fin_merge_ti (*x*#*xs*) *y* =
  (*case y of* [] $\Rightarrow$ (*x*#*xs*)
    | (*z*#*zs*) $\Rightarrow$ (*x*@*z*) # (*fin_merge_ti xs zs*))

constdefs
inf_merge_ti :: '*a istream* $\Rightarrow$ '*a istream* $\Rightarrow$ '*a istream*
inf_merge_ti x y
  $\equiv$
  λ *i*. (*x i*)@(*y i*)

primrec
fin_last_ti s 0 = *hd s*
fin_last_ti s (*Suc i*) =
  (*if s*!(*Suc i*) $\neq$ []
    then s!(*Suc i*)
    else fin_last_ti s i)

primrec
inf_last_ti s 0 = *s 0*
inf_last_ti s (*Suc i*) =
  (*if s* (*Suc i*) $\neq$ []
    then s (*Suc i*)
    else inf_last_ti s i)

lemma *inf_last_ti_nonempty_k*:
inf_last_ti dt t $\neq$ []
  $\Longrightarrow$ *inf_last_ti dt* (*t* + *k*) $\neq$ []
  by (*induct k, auto*)

lemma *inf_last_ti_nonempty*:
s t $\neq$ [] $\Longrightarrow$ *inf_last_ti s* (*t* + *k*) $\neq$ []
  apply (*induct k, auto*)
  apply (*induct t, auto*)
  done

Lemmas for Concatenation Operator

lemma *fin_length_append*:
fin_length (*x*@*y*) = (*fin_length x*) + (*fin_length y*)
  apply (*induct x*)
  apply *auto*
  done

lemma *fin_append_Nil*:
fin_inf_append [] *z* = *z*
  by (*simp add: fin_inf_append_def*)

lemma *correct_fin_inf_append1*:
s1 = *fin_inf_append* [*x*] *s* $\Longrightarrow$ *s1* (*Suc i*) = *s i*

by (*simp add: fin_inf_append_def*)

lemma *correct_fin_inf_append2*:
fin_inf_append [*x*] *s* (*Suc i*) = *s i*
  by (*simp add: fin_inf_append_def*)

lemma *fin_append_com_Nil1*:
fin_inf_append [] (*fin_inf_append y z*)
 = *fin_inf_append* ([] @ *y*) *z*
  by (*simp add: fin_append_Nil*)

lemma *fin_append_com_Nil2*:
fin_inf_append x (*fin_inf_append* [] *z*) = *fin_inf_append* (*x* @ []) *z*
  by (*simp add: fin_append_Nil*)

lemma *fin_append_com_i*:
fin_inf_append x (*fin_inf_append y z*) *i* = *fin_inf_append* (*x* @ *y*) *z i*
  apply (*case_tac x*)
  apply (*simp add: fin_append_com_Nil1*)
  apply (*case_tac y*)
  apply (*simp add: fin_append_com_Nil2*)
  apply (*simp add: fin_inf_append_def*)
  apply *auto*
  apply (*simp add: list_nth_append2*)
  apply (*simp add: list_nth_append3*)
  apply (*simp add: list_nth_append9*)
  apply (*simp add: list_nth_append10*)
  done

Lemmas for Operators ts and msg

lemma *ts_msg1*:
ts p $\Longrightarrow$ *msg 1 p*
  by (*simp add: ts_def msg_def*)

lemma *msg_time_interval*:
$\forall$ *s*. *msg 1 s* $\Longrightarrow$
($\forall$ *t x*. *s t* = *x* $\land$ *x* $\neq$ [] $\Longrightarrow$
($\exists$ *a*. *s t* = [*a*] $\land$ *a* = *hd x*))
  by *simp*

lemma *ts_inf_tl*:
ts x $\Longrightarrow$ *ts* (*inf_tl x*)
  by (*simp add: ts_def inf_tl_def*)

lemma *ts_Least_0*:
ts x $\Longrightarrow$ (*LEAST i*. (*x i*) $\neq$ []) = (0::*nat*)
  apply (*simp add: ts_def*)
  apply (*erule_tac x*=0::*nat* **in** *allE*)
  apply (*subgoal_tac x 0* $\neq$ [])
    prefer *2*
    apply (*simp add: list_length_hint3*)
  apply (*simp add: Least_def*)

apply *auto*
done

lemma *lambda_Suc*:
$(\lambda i.\ x\ (Suc\ i))\ i = x\ (Suc\ i)$
  by *auto*

lemma *inf_tl_Suc*:
inf_tl x i $= x\ (Suc\ i)$
  by (*simp add: inf_tl_def*)

lemma *ts_length_hint3*:
ts x $\implies x\ i \neq$ []
  apply (*simp add: ts_def*)
  apply (*erule_tac x=i in allE*)
  apply (*simp add: list_length_hint3*)
  done

lemma *ts_Least_Suc0*:
ts x $\implies (LEAST\ i.\ x\ (Suc\ i) \neq$ []$) = 0$
  apply (*subgoal_tac* $\forall\ j.\ x\ j \neq$ [])
    prefer *2*
    apply (*simp add: ts_length_hint3*)
  apply *auto*
  apply (*simp add: ts_def*)
  apply (*erule_tac x=0::nat in allE*)
  apply (*simp add: Least_def*)
  apply *auto*
  done

lemma *ts_inf_make_untimed_hd*:
$\bigwedge x.\ ts\ x \implies (inf\_make\_untimed\ x)\ i = hd\ (x\ i)$
  apply (*simp add: inf_make_untimed_def*)
  apply (*induct i*)
  apply *auto*
    apply (*simp add: ts_Least_0*)
  apply (*simp add: ts_def list_length_hint3*)
  apply (*simp add: ts_def*)
  apply (*subgoal_tac ts (inf_tl x)*)
    prefer *2*
    apply (*simp add: ts_inf_tl*)
  apply (*case_tac i*)
  apply *simp*
  apply (*simp add: inf_tl_Suc*)
  apply *atomize*
  apply (*erule_tac x=inf_tl x in allE*)
  apply *clarify*
  apply (*subgoal_tac inf_tl x (Suc nat) = x (Suc (Suc nat))*)
    prefer *2*
    apply (*simp add: inf_tl_Suc*)
  apply *simp*
  done

A.1.1. Lemmas about inf_truncate

lemma *inf_truncate_nonempty*:
$[\![z\ i \neq []]\!] \implies$ *inf_truncate* $z\ i \neq []$
  by (*induct i, auto*)

lemma *concat_inf_truncate_nonempty*:
$[\![z\ i \neq []]\!] \implies$ *concat* (*inf_truncate* $z\ i$) $\neq []$
  by (*induct i, auto*)

lemma *concat_inf_truncate_nonempty_el*:
$[\![z\ i = [a]]\!] \implies$ *concat* (*inf_truncate* $z\ i$) $\neq []$
  by (*induct i, auto*)

lemma *inf_truncate_append*:
(*inf_truncate* $z\ i$ @ [z (*Suc i*)]) = *inf_truncate* z (*Suc i*)
  by (*induct i, simp+*)

A.1.2. Lemmas about fin_make_untimed

lemma *fin_make_untimed_append*:
fin_make_untimed $x \neq [] \implies$ *fin_make_untimed* (x @ y) $\neq []$
  by (*simp add: fin_make_untimed_def*)

lemma *fin_make_untimed_inf_truncate_Nonempty*:
$\bigwedge z.\ [\![z\ k \neq [];\ k \leq i]\!]$
  $\implies$ *fin_make_untimed* (*inf_truncate* $z\ i$) $\neq []$
  apply (*simp add: fin_make_untimed_def*)
  apply (*induct i*)
  apply *simp*
  apply *atomize*
  apply (*erule_tac x=z in allE*)
  apply *simp*
  apply (*case_tac z* (*Suc i*) = [])
  apply *auto*
  apply (*induct k*)
  apply *simp*
  apply *atomize*
  apply (*erule_tac x=i in allE*)
  apply (*erule_tac x=z in allE*)
  apply *simp*
  apply (*subgoal_tac k = i*)
    prefer *2*
    apply *simp*
  apply *simp*
  done

lemma *last_fin_make_untimed_append*:
last (*fin_make_untimed* (z @ [[a]])) = a
  by (*simp add: fin_make_untimed_def*)

lemma *last_fin_make_untimed_inf_truncate*:
$[\![z\ i = [a]]\!] \implies$
last (*fin_make_untimed* (*inf_truncate* $z\ i$)) = a

apply (*simp add: fin_make_untimed_def*)
apply (*induct i*)
apply *auto*
done

lemma *fin_make_untimed_append_empty*:
fin_make_untimed (*z* @ [[]]) = *fin_make_untimed z*
  by (*simp add: fin_make_untimed_def*)

lemma *fin_make_untimed_inf_truncate_append_a*:
fin_make_untimed (*inf_truncate z i* @ [[*a*]]) !
  (*length* (*fin_make_untimed* (*inf_truncate z i* @ [[*a*]]))) − *Suc 0*) = *a*
  by (*simp add: fin_make_untimed_def*)

lemma *fin_make_untimed_inf_truncate_Nonempty_all*:
⟦ *z k* ≠ [] ⟧
  ⟹ ∀ *i. k* ≤ *i* ⟶ *fin_make_untimed* (*inf_truncate z i*) ≠ []
  by (*simp add: fin_make_untimed_inf_truncate_Nonempty*)

lemma *fin_make_untimed_inf_truncate_Nonempty_all0*:
⟦ *z 0* ≠ [] ⟧
  ⟹ ∀ *i. fin_make_untimed* (*inf_truncate z i*) ≠ []
  by (*simp add: fin_make_untimed_inf_truncate_Nonempty*)

lemma *fin_make_untimed_inf_truncate_Nonempty_all0a*:
⟦ *z 0* = [*a*] ⟧
  ⟹ ∀ *i. fin_make_untimed* (*inf_truncate z i*) ≠ []
  by (*simp add: fin_make_untimed_inf_truncate_Nonempty_all0*)

lemma *fin_make_untimed_inf_truncate_Nonempty_all_app*:
⟦ *z 0* = [*a*] ⟧
  ⟹ ∀ *i. fin_make_untimed* (*inf_truncate z i* @ [*z* (*Suc i*)]) ≠ []
  apply *clarify*
  apply (*subgoal_tac fin_make_untimed* (*inf_truncate z i*) ≠ [])
    prefer *2*
  apply (*simp add: fin_make_untimed_inf_truncate_Nonempty_all0a*)
  apply (*simp add: fin_make_untimed_append*)
  done

lemma *fin_make_untimed_nth_length*:
⟦ *z i* = [*a*] ⟧ ⟹
fin_make_untimed (*inf_truncate z i*) !
  (*length* (*fin_make_untimed* (*inf_truncate z i*)) − *Suc 0*)
= *a*
  apply (*subgoal_tac last* (*fin_make_untimed* (*inf_truncate z i*)) = *a*)
    prefer *2*
    apply (*simp add: last_fin_make_untimed_inf_truncate*)
  apply (*simp add: fin_make_untimed_def*)
  apply (*subgoal_tac concat* (*inf_truncate z i*) ≠ [])
    prefer *2*
    apply (*erule concat_inf_truncate_nonempty_el*)
  apply (*simp add: last_nth_length*)
  done

A.1.3. Lemmas about inf_disj and inf_disjS

lemma *inf_disj_index*:
$[\![\ $ *inf_disj n nS*; $\ nS\ k\ t \neq [];\ \ k < n\ ]\!]$
$\quad \Longrightarrow (SOME\ i.\ i < n \wedge\ nS\ i\ t \neq []) = k$
apply (*simp add: inf_disj_def*)
apply (*erule_tac x=t* **in** *allE*)
apply (*erule_tac x=k* **in** *allE*)
apply *auto*
done

lemma *inf_disjS_index*:
$[\![\ $ *inf_disjS IdSet nS*; *k:IdSet*; $\ nS\ k\ t \neq []\ ]\!]$
$\quad \Longrightarrow (SOME\ i.\ (i:IdSet) \wedge\ nSend\ i\ t \neq []) = k$
apply (*simp add: inf_disjS_def*)
apply (*erule_tac x=t* **in** *allE*)
apply (*erule_tac x=k* **in** *allE*)
apply *auto*
done
end

A.2. Theory join_ti – Concatenation of time intervals

theory *join_ti = Main + stream + arith_hints*:

Definition and lemmas for infinite timed streams:

consts
  join_ti ::*'a istream* $\Rightarrow$ *nat* $\Rightarrow$ *nat* $\Rightarrow$ *'a list*
primrec
join_ti_0:
  join_ti s x 0 = s x
join_ti_Suc:
  join_ti s x (Suc i) = (join_ti s x i) @ *(s (x + (Suc i)))*

lemma *join_ti_hint1*:
join_ti s x (Suc i) = $[]$ $\Longrightarrow$ *join_ti s x i* = $[]$
  by *auto*

lemma *join_ti_hint2*:
join_ti s x (Suc i) = $[]$ $\Longrightarrow$ *s (x + (Suc i))* = $[]$
  by *auto*

lemma *join_ti_hint3*:
join_ti s x (Suc i) = $[]$ $\Longrightarrow$ *s (x + i)* = $[]$
  by (*induct i, auto*)

lemma *join_ti_empty_join*:
$[\![\ i \leq n;\ join\_ti\ s\ x\ n = []\ ]\!] \Longrightarrow s\ (x{+}i) = []$
  apply (*induct n*)
  apply *auto*
  apply (*case_tac i = Suc n*)
  apply *simp+*
  done

lemma *join_ti_empty_ti*:
$[\![\ \forall\ i \leq n.\ \ s\ (x{+}i) = []\ ]\!] \Longrightarrow join\_ti\ s\ x\ n = []$

by (*induct n, auto*)

lemma *join_ti_1nempty*:
⟦ ∀ *i*. *0* < *i* ∧ *i* < *Suc n* ⟶ *s* (*x*+*i*) = [] ⟧
⟹ *join_ti s x n = s x*
  by (*induct n, auto*)

  Definition and lemmas for finite timed streams:

consts
  fin_join_ti ::*'a fstream* ⇒ *nat* ⇒ *nat* ⇒ *'a list*
primrec
fin_join_ti_0:
  fin_join_ti s x 0 = nth s x
fin_join_ti_Suc:
  fin_join_ti s x (Suc i) = (fin_join_ti s x i) @ (nth s (x + (Suc i)))

lemma *fin_join_ti_hint1*:
fin_join_ti s x (Suc i) = [] ⟹ fin_join_ti s x i = []
  by *auto*

lemma *fin_join_ti_hint2*:
fin_join_ti s x (Suc i) = [] ⟹ nth s (x + (Suc i)) = []
  by *auto*

lemma *fin_join_ti_hint3*:
fin_join_ti s x (Suc i) = [] ⟹ nth s (x + i) = []
  by (*induct i, auto*)

lemma *fin_join_ti_empty_join*:
⟦ *i* ≤ *n*; *fin_join_ti s x n* = [] ⟧ ⟹ *nth s (x+i)* = []
  apply (*induct n*)
  apply *auto*
  apply (*case_tac i = Suc n*)
  apply *simp*+
  done

lemma *fin_join_ti_empty_ti*:
⟦ ∀ *i* ≤ *n*. *nth s (x+i)* = [] ⟧ ⟹ *fin_join_ti s x n* = []
  by (*induct n, auto*)

lemma *fin_join_ti_1nempty*:
⟦ ∀ *i*. *0* < *i* ∧ *i* < *Suc n* ⟶ *nth s (x+i)* = [] ⟧
⟹ *fin_join_ti s x n = nth s x*
  by (*induct n, auto*)
end

A.3. Changing Time Granularity

theory *fin_time_raster = Main + stream + join_ti*:

Split time intervals

consts
  fin_split_time ::*'a fstream* ⇒ *nat* ⇒ *'a fstream*
primrec
  fin_split_time [] *n* = []

```
fin_split_time (x#xs) n =
  (case n of
    0 ⇒ []
  | Suc i ⇒ x # (replicate i []) @ (fin_split_time xs n) )
```

lemma *fin_split_time1*:
fin_split_time s (Suc 0) = s
  by (*induct s, auto*)

lemma *fin_split_time1t*:
⋀ *t. (fin_split_time s 1)!t = s!t*
  by (*simp add: fin_split_time1*)

lemma *fin_split_time_drop*:
⟦ *(0::nat) < n* ⟧ ⟹
drop n (fin_split_time (a # l) n) = fin_split_time l n
  by (*induct n, auto*)

Join time intervals

lemma *fin_join_time_drop_hint*:
n ≤ length s ∧ 0 < n ∧ s ≠ [] ⟶ length (drop n s) < length s
  by *auto*

consts
  fin_join_time :: 'a fstream × nat ⇒ 'a fstream
recdef
fin_join_time measure ((λ(s,n). length s))
fin_join_time (s, 0) = []
fin_join_time (s, Suc i) =
  (*if s = []*
    then []
    else
      (*if (length s) < (Suc i)*
        then s
        else (concat (take (Suc i) s))
            # fin_join_time (drop (Suc i) s, (Suc i))))

lemma *fin_join_time1*:
fin_join_time(s, Suc 0) = s
  by (*induct_tac s, simp+*)

lemma *fin_join_time1t*:
∀ *t. fin_join_time (s, Suc 0)! t = s! t*
  by (*simp only: fin_join_time1, simp*)

lemma *fin_join_time2*:
fin_join_time([], i) = []
  by (*induct_tac i, simp+*)

Duality of the split and the join operators

lemma *fin_split_time_n1*:

⟦ *s* ≠ []; *n* = *Suc i* ⟧ ⟹ *take n* (*fin_split_time s n*) = (*hd s*) # (*replicate i* [])
  by (*induct s, auto*)

lemma *fin_split_time_n2*:
⟦ *s* ≠ []; *0* < *n* ⟧ ⟹ *concat* (*take n* (*fin_split_time s n*)) = *hd s*
  apply (*induct s*)
  apply *auto*
  apply (*case_tac n*)
  apply *simp*
  apply (*rename_tac a s i*)
  apply (*subgoal_tac concat* (*a* # (*replicate i* [])) = *a*)
  apply (*subgoal_tac*
  take n (*a* # *replicate i* [] @ *fin_split_time s n*)
    = *a* # *replicate i* [])
  apply *auto*
  apply (*simp add: set_replicate_conv_if*)
  apply (*simp split: split_if_asm*)
  done

lemma *fin_join_split*:
⟦ (*0::nat*) < *n* ⟧ ⟹ *fin_join_time* (*fin_split_time s n, n*) = *s*
  apply (*induct_tac s*)
  apply (*simp add: fin_join_time2*)
  apply (*rename_tac a l*)
  apply (*subgoal_tac concat* (*take n* (*fin_split_time* (*a* # *l*) *n*)) = *a*)
    prefer *2*
    apply (*erule_tac V* = *fin_join_time* (*fin_split_time l n, n*) = *l* **in** *thin_rl*)
    apply (*subgoal_tac concat* (*take n* (*fin_split_time* (*a* # *l*) *n*)) = *hd* (*a* # *l*))
      prefer *2*
      apply (*subgoal_tac a#l* ≠ [])
        prefer *2*
        apply *simp*
      apply (*erule fin_split_time_n2, assumption*)
    apply *simp*
  apply (*subgoal_tac drop n* (*fin_split_time* (*a* # *l*) *n*) = *fin_split_time l n*)
    prefer *2*
    apply (*erule fin_split_time_drop*)
  apply (*subgoal_tac*
  fin_join_time (*fin_split_time* (*a* # *l*) *n, n*) =
  concat (*take n* (*fin_split_time* (*a* # *l*) *n*))
    # *fin_join_time* ((*drop n* (*fin_split_time* (*a* # *l*) *n*)), *n*))
  apply *auto*
  apply (*case_tac n, simp+*)+
  done
end

theory *time_raster* = *Main* + *stream* + *join_ti*:

Split time intervals

constdefs
  split_time ::'*a istream* ⇒ *nat* ⇒ '*a istream*
  split_time s n t ≡
  (*if* (*t mod n* = *0*)
    then s (*t div n*)
    else [])

lemma *split_time1t*:
$\forall\ t.\ split\_time\ s\ 1\ t = s\ t$
  by (*simp add: split_time_def*)

lemma *split_time1*:
split_time s 1 = s
  by (*simp add: expand_fun_eq split_time_def*)

lemma *split_time_mod*:
$t\ mod\ n \neq 0 \implies split\_time\ s\ n\ t = []$
  by (*simp add: split_time_def*)

lemma *split_time_nempty*:
$[\![\ 0 < n\ ]\!] \implies split\_time\ s\ n\ (n * t) = s\ t$
  by (*simp add: split_time_def*)

lemma *split_time_nempty_Suc*:
$[\![\ 0 < n\ ]\!] \implies$
*split_time s (Suc n) ((Suc n) * t) = split_time s n (n * t)*
  apply (*subgoal_tac split_time s (Suc n) ((Suc n) * t) = s t*)
  apply (*simp add: split_time_nempty*)
  apply (*subgoal_tac 0 < Suc n*)
  apply (*erule split_time_nempty*)
  apply *simp*
  done

lemma *split_time_empty*:
$[\![\ i < n;\ 0 < i\ ]\!] \implies split\_time\ s\ n\ (n * t + i) = []$
  apply (*simp add: split_time_def*)
  apply (*subgoal_tac 0 < (n * t + i) mod n*)
  apply *simp*
  apply (*erule arith_mod_nzero, assumption*)
  done

lemma *split_time_empty_Suc*:
$[\![\ i < n;\ 0 < i\ ]\!] \implies$
split_time s (Suc n) ((Suc n) t + i) = split_time s n (n * t + i)*
  apply (*subgoal_tac 0 < Suc n*)
  apply (*subgoal_tac i < Suc n*)
  apply (*subgoal_tac split_time s (Suc n) (Suc n * t + i) = []*)
  apply (*simp add: split_time_empty*)
  apply (*erule split_time_empty, assumption*)
  apply *simp+*
  done

lemma *split_time_hint1*:
$n = Suc\ m \implies split\_time\ s\ (Suc\ n)\ (i + n * i + n) = []$
  apply (*subgoal_tac i + n * i + n = (Suc n) * i + n*)
    prefer *2*
    apply *simp*
  apply (*subgoal_tac n < Suc n*)
    prefer *2*
    apply *simp*
  apply (*subgoal_tac split_time s (Suc n) (Suc n * i + n) = []*)
    prefer *2*
    apply (*rule split_time_empty*)
  apply *auto*

done

Join time intervals

constdefs
  join_time :: $'a$ *istream* $\Rightarrow$ *nat* $\Rightarrow$ $'a$ *istream*
  join_time s n t $\equiv$
  (*case n of*
    $0 \Rightarrow []$
  $|(Suc\ i) \Rightarrow$ *join_ti s* $(n*t)\ i)$

lemma *join_time1t*:
$\forall\ t.\ join\_time\ s\ (1::nat)\ t = s\ t$
  by (*simp add: join_time_def*)

lemma *join_time1*:
join_time s 1 = s
  apply (*simp add: expand_fun_eq*)
  apply (*simp add: join_time_def*)
  done

lemma *join_time_empty1*:
$[\![\ i < n;\ join\_time\ s\ n\ t = [\,]\ ]\!] \Longrightarrow s\ (n*t + i) = [\,]$
  apply (*simp add: join_time_def*)
  apply (*case_tac n*)
  apply *simp+*
  apply *clarify*
  apply (*subgoal_tac i $\leq$ nat*)
    prefer *2*
    apply *simp*
  apply (*simp add: join_ti_empty_join*)
  done

Duality of the split and the join operators

lemma *join_split_i*:
$[\![\ 0 < n\ ]\!] \Longrightarrow join\_time\ (split\_time\ s\ n)\ n\ i = s\ i$
  apply (*simp add: join_time_def*)
  apply (*case_tac n*)
  apply *simp+*
  apply *clarify*
  apply (*rename_tac n*)
  apply (*subgoal_tac i + n $*$ i = (Suc n) $*$ i*)
    prefer *2*
    apply *simp*
  apply (*subgoal_tac 0 < Suc n*)
    prefer *2*
    apply *simp*
  apply (*subgoal_tac join_ti (split_time s (Suc n)) ((Suc n) $*$ i) n = s i*)
  apply *simp*
  apply (*subgoal_tac (split_time s (Suc n)) (Suc n $*$ i) = s i*)
    prefer *2*
    apply (*erule split_time_nempty*)
  apply (*subgoal_tac
  $\forall\ j.\ 0 < j \wedge j < Suc\ n \longrightarrow split\_time\ s\ (Suc\ n)\ (Suc\ n\ * i + j) = [\,])$*

    prefer *2*
    apply *clarify*
    apply (*erule split_time_empty, assumption*)
    apply (*simp add: join_ti_1nempty*)
    done

lemma *join_split*:
⟦ *0 < n* ⟧ ⟹ *join_time* (*split_time s n*) *n* = *s*
  by (*simp add: expand_fun_eq join_split_i*)
end

A.4. Theory ArithExtras.thy

theory *ArithExtras = Main + Nat_Infinity*:

consts
   nat2inat :: nat list ⇒ inat list
primrec
  nat2inat [] = []
  nat2inat (*x#xs*) = (*Fin x*) # (*nat2inat xs*)

end

A.5. Theory ListExtras.thy

theory *ListExtras = Main + List_Prefix*:
constdefs
   disjoint :: 'a list ⇒ 'a list ⇒ bool
  disjoint x y ≡ (*set x*) ∩ (*set y*) = {}

lemma *set_inter_mem*:
⟦*x mem l1; x mem l2* ⟧ ⟹ *set l1* ∩ *set l2* ≠ {}
  apply (*induct l1*)
  apply *simp*
  apply (*simp split add: split_if_asm*)
  apply *auto*
  apply (*induct l2*)
  apply *simp*
  apply (*simp split add: split_if_asm*)
  apply *auto*
  done

lemma *mem_notdisjoint*:
⟦*x mem l1; x mem l2* ⟧ ⟹ ¬ *disjoint l1 l2*
  apply (*simp add: disjoint_def*)
  apply (*subgoal_tac* ⟦*x mem l1; x mem l2* ⟧ ⟹ *set l1* ∩ *set l2* ≠ {})
  apply *simp*
  apply (*simp add: set_inter_mem*)
  done
lemma *Add_Less: 0 < b* ⟹ (*Suc a − b < Suc a*) = *True*
  by *arith*

lemma *list_length_hint1: l ~= []* ⟹ *0 < length l*
  by *simp*

lemma *list_length_hint1a*: $\llbracket l \sim= [] \rrbracket \Longrightarrow 0 < length\ l$
  by *simp*

lemma *list_length_hint2*: $length\ x = Suc\ 0 \Longrightarrow [hd\ x] = x$
  by (*induct x, auto*)

lemma *list_length_hint2a*: $length\ l = Suc\ 0 \Longrightarrow tl\ l = []$
  by (*induct l, auto*)

lemma *list_length_hint3*: $length\ l = Suc\ 0 \Longrightarrow l \neq []$
  by (*induct l, auto*)

lemma *list_length_hint4*:
$\llbracket length\ x \leq Suc\ 0;\ x \neq [] \rrbracket \Longrightarrow length\ x = Suc\ 0$
  by (*induct x, auto*)

lemma *list_nth_append1*:
$\llbracket i < length\ x \rrbracket \Longrightarrow (b\ \#\ x)\ !\ i = (b\ \#\ x\ @\ y)\ !\ i$
  apply (*case_tac i, auto*)
  apply (*simp add: nth_append*)
  done

lemma *list_nth_append2*:
$\llbracket i < Suc\ (length\ x) \rrbracket \Longrightarrow (b\ \#\ x)\ !\ i = (b\ \#\ x\ @\ a\ \#\ y)\ !\ i$
  apply (*case_tac i, auto*)
  apply (*simp add: nth_append*)
  done

lemma *list_nth_append3*:
$\llbracket \neg\ i < Suc\ (length\ x);\ i - Suc\ (length\ x) < Suc\ (length\ y) \rrbracket$
$\Longrightarrow (a\ \#\ y)\ !\ (i - Suc\ (length\ x)) = (b\ \#\ x\ @\ a\ \#\ y)\ !\ i$
  apply (*case_tac i, auto*)
  apply (*simp add: nth_append*)
  done

lemma *list_nth_append4a*:
$\llbracket i < Suc\ (length\ x + length\ y);\ \neg\ i - Suc\ (length\ x) < Suc\ (length\ y) \rrbracket$
$\Longrightarrow False$
  by *arith*

lemma *list_nth_append4*:
$\llbracket i - length\ x < Suc\ (length\ y);\ \neg\ i - Suc\ (length\ x) < Suc\ (length\ y) \rrbracket$
$\Longrightarrow \neg\ i < Suc\ (length\ x + length\ y)$
  by *arith*

lemma *list_nth_append5*:
$\llbracket \neg\ i < Suc\ (length\ x);\ i - Suc\ (length\ x) < Suc\ (length\ y) \rrbracket$
$\Longrightarrow (a\ \#\ y)\ !\ (i - Suc\ (length\ x)) = (aa\ \#\ x\ @\ a\ \#\ y)\ !\ i$
  apply (*case_tac i, auto*)
  apply (*simp add: nth_append*)
  done

lemma *list_nth_append6*:
$\llbracket \neg\ i - length\ x < Suc\ (length\ y);\ \neg\ i - Suc\ (length\ x) < Suc\ (length\ y) \rrbracket$
$\Longrightarrow \neg\ i < Suc\ (length\ x + length\ y)$
  by *arith*

lemma *list_nth_append6a*:
$[\![i < Suc \ (length \ x + length \ y); \ \neg \ i - length \ x < Suc \ (length \ y)]\!]$
$\implies False$
by *arith*

lemma *list_nth_append7*:
$[\![i - length \ x < Suc \ (length \ y); \ \ i - Suc \ (length \ x) < Suc \ (length \ y)]\!]$
$\implies i < Suc \ (Suc \ (length \ x + length \ y))$
apply *arith*
done

lemma *list_nth_append8*:
$[\![\ \neg \ i < Suc \ (length \ x + length \ y); \ \ i < Suc \ (Suc \ (length \ x + length \ y))]\!]$
$\implies \ i = Suc \ (length \ x + length \ y)$
by *arith*

lemma *list_nth_append9*:
$[\![\ i - Suc \ (length \ x) < Suc \ (length \ y)]\!]$
$\implies \ i < Suc \ (Suc \ (length \ x + length \ y))$
by *arith*

lemma *list_nth_append10*:
$[\![\neg \ i < Suc \ (length \ x); \ \neg \ i - Suc \ (length \ x) < Suc \ (length \ y)]\!]$
$\implies \neg \ i < Suc \ (Suc \ (length \ x + length \ y))$
by *arith*
end

A.6. Auxiliary Arithmetic Lemmas

theory *arith_hints* = *Main*:

lemma *arith_mod_neq*:
$[\![\ a \ mod \ n \neq b \ mod \ n \ ]\!] \implies a \neq b$
by *auto*

lemma *arith_mod_nzero*:
$[\![\ i < n; \ (0{::}nat) < i]\!] \implies 0 < (n * t + i) \ mod \ n$
  apply (*subgoal_tac* $(i + n * t) \ mod \ n = i$)
    prefer *2*
    apply (*simp add*: *mod_mult_self2*)
  apply (*subgoal_tac* $n * t + i = i + n * t$)
    prefer *2*
    apply *simp*
  apply (*simp* (*no_asm_simp*))
  done

lemma *arith_20*:
$\bigwedge \ i. \ [\![\ (i{::}nat) < n; \ 0 < i]\!] \implies i + n * t \neq n * q$
  apply (*subgoal_tac* $(i + n * t) \ mod \ n = i$)
    prefer *2*
    apply (*simp add*: *mod_mult_self2*)
  apply (*subgoal_tac* $(n * q) \ mod \ n = 0$)
    prefer *2*

    apply *simp*
    apply (*subgoal_tac* $(i + n * t)$ *mod* $n \neq (n * q)$ *mod* n)
      prefer *2*
      apply *simp*
    thm *arith_mod_neq*
    apply (*erule arith_mod_neq*)
    done

lemma *arith_21*:
$\bigwedge i.$ ⟦ $(i{::}nat) < n;\ 0 < i$⟧ $\Longrightarrow n * t + i \neq n * q$
    apply (*subgoal_tac* $n * t + i = i + n * t$)
      prefer *2*
      apply *simp*
    apply (*subgoal_tac* $i + n * t \neq n * q$)
    apply *simp*
    apply (*erule arith_20, assumption*)
    done

lemma *arith_22*:
$\bigwedge i.$ ⟦ $(i{::}nat) < n;\ 0 < i$ ⟧ $\Longrightarrow n + n * t + i \neq n * qc$
    apply (*subgoal_tac* $n + n * t + i = n * (Suc\ t) + i$)
      prefer *2*
      apply *simp*
    apply (*subgoal_tac* $\bigwedge i.$ ⟦ $(i{::}nat) < n;\ 0 < i$ ⟧ $\Longrightarrow n*(Suc\ t) + i \neq n * qc$)
      prefer *2*
      apply (*rule arith_21, assumption+*)
    apply *simp*
    done

lemma *split_time_0arith_mod1*:
$(t + n * t)$ *mod* $Suc\ n = 0$
    apply (*subgoal_tac* $((Suc\ n) * t)$ *mod* $Suc\ n = 0$)
      prefer *2*
      apply (*rule mod_mult_self1_is_0*)
    apply *simp*
    done

lemma *split_time_0arith_mod2*:
$Suc\ (n + (t + n * t))$ *mod* $Suc\ n = 0$
    apply (*subgoal_tac* $((Suc\ n) * (Suc\ t))$ *mod* $Suc\ n = 0$)
      prefer *2*
      apply (*rule mod_mult_self1_is_0*)
    apply *simp*
    done

lemma *split_time_0arith0*:
$t + n * t = (Suc\ n) * t$
    by *auto*

lemma *split_time_0arith1*:
⟦ $Suc\ n * t = Suc\ n * q$⟧ $\Longrightarrow t = q$

    apply (*subgoal_tac* $Suc\ n * t = Suc\ n * q = (t = q \mid (Suc\ n) = (0{::}nat))$)
      prefer *2*
      thm *mult_cancel1*

```
    apply (rule mult_cancel1)
  apply simp
  done

lemma split_time_0arith2:
(t::nat) + (n::nat) * t = (q::nat) + n * q ⟹ t = q
  quickcheck
  apply (subgoaLtac t + n * t = (Suc n) * t)
    prefer 2
    apply (rule split_time_0arith0)
  apply (subgoaLtac q + n * q = (Suc n) * q)
    prefer 2
    apply (rule split_time_0arith0)
  apply (subgoaLtac Suc n * t = Suc n * q)
  apply (erule split_time_0arith1)
  apply (simp (no_asm))
  done
end
```

Appendix B.

Isabelle/HOL Specifications and Proofs

B.1. Steam Boiler System Specification

theory *SteamBoiler* = *Main* + *stream* + *Bit*:

constdefs *ControlSystem* :: *nat istream* $\Rightarrow$ *bool*
ControlSystem s $\equiv$
$(ts\ s)\ \wedge$
$(\forall\ (j::nat).\ (200::nat) \leq hd\ (s\ j) \wedge hd\ (s\ j) \leq (800::\ nat))$

constdefs
SteamBoiler :: *bit istream* $\Rightarrow$ *nat istream* $\Rightarrow$ *nat istream* $\Rightarrow$ *bool*
SteamBoiler x s y $\equiv$
ts x
$\longrightarrow$
$((ts\ y) \wedge (ts\ s) \wedge (y = s) \wedge$
$(hd\ (s\ 0) = (500::nat)) \wedge$
$(\forall\ (j::nat).\ (\exists\ (r::nat).$
$\qquad (0::nat) < r \wedge r \leq (10::nat) \wedge$
$\qquad hd\ (s\ (Suc\ j)) =$
$\qquad\quad (if\ hd\ (x\ j) = Zero$
$\qquad\quad then\ (hd\ (s\ j)) - r$
$\qquad\quad else\ (hd\ (s\ j)) + r))\ ))$

constdefs
Converter :: *bit istream* $\Rightarrow$ *bit istream* $\Rightarrow$ *bool*
Converter z x
$\equiv$
$(ts\ x)$
$\wedge$
$(\forall\ (t::nat).$
$\quad hd\ (x\ t) =$
$\qquad (if\ (fin\_make\_untimed\ (inf\_truncate\ z\ t) = [])$
$\qquad then$
$\qquad\quad Zero$
$\qquad else$
$\qquad\quad (fin\_make\_untimed\ (inf\_truncate\ z\ t))\ !$
$\qquad\qquad ((length\ (fin\_make\_untimed\ (inf\_truncate\ z\ t))) - (1::nat))$
$\quad))$

constdefs
Controller_L ::
$\quad$ *nat istream* $\Rightarrow$ *bit iustream* $\Rightarrow$ *bit iustream* $\Rightarrow$ *bit istream* $\Rightarrow$ *bool*
Controller_L y lIn lOut z

$\equiv$
$(z\ 0\ =\ [Zero])$
$\wedge$
$(\forall\ (t::nat).$
$(\ if\ (lIn\ t)\ =\ Zero$
$\quad then\ (\ if\ 300\ <\ hd\ (y\ t)$
$\qquad\qquad then\ \ (z\ t)\ =\ []\quad \wedge\ (lOut\ t)\ =\ Zero$
$\qquad\qquad else\ \ (z\ t)\ =\ [One]\ \wedge\ (lOut\ t)\ =\ One$
$\qquad\qquad)$
$\quad else\ (\ if\ \ hd\ (y\ t)\ <\ 700$
$\qquad\qquad then\ \ (z\ t)\ =\ []\quad \wedge\ (lOut\ t)\ =\ One$
$\qquad\qquad else\ \ (z\ t)\ =\ [Zero]\ \wedge\ (lOut\ t)\ =\ Zero\ )\ ))$

constdefs
  Controller :: nat istream $\Rightarrow$ bit istream $\Rightarrow$ bool
  Controller y z
$\equiv$
$(ts\ y)$
$\longrightarrow$
$(\exists\ l.\ Controller\_L\ y\ (fin\_inf\_append\ [Zero]\ l)\ l\ z)$

constdefs
  ControlSystemArch :: nat istream $\Rightarrow$ bool
  ControlSystemArch s
$\equiv$
$\exists\ x\ z\ ::\ bit\ istream.\ \exists\ y\ ::\ nat\ istream.$
  $(\ SteamBoiler\ x\ s\ y\ \wedge\ Controller\ y\ z\ \wedge\ Converter\ z\ x\ )$

end

B.2. Proof of the Steam Boiler System Properties

theory *SteamBoiler_proof = Main + SteamBoiler:*

B.2.1. Properties of Controller Component

lemma *L1_Controller:*
$[\![\ Controller\_L\ y\ (fin\_inf\_append\ [Zero]\ l)\ l\ z;$
  $l\ t\ \neq\ Zero\ ]\!]$
$\Longrightarrow last\ (fin\_make\_untimed\ (inf\_truncate\ z\ t))\ =\ One$
apply *(induct t)*
apply *(simp add: Controller_L_def)*
apply *clarify*
apply *(erule_tac x=0 in allE)*
apply *(simp split add: split_if_asm)*
apply *simp*
apply *(subgoal_tac*
  Controller_L y (fin_inf_append [Zero] l) l z)
  prefer *2*
  apply *simp*
apply *(simp add: Controller_L_def)*
apply *clarify*
apply *(erule_tac x=Suc t in allE)*

apply (*erule_tac x=t* **in** *allE*)
apply (*simp split add: split_if_asm*)
apply *clarify*
apply (*simp add: fin_make_untimed_append_empty fin_inf_append_def*)
apply (*simp add: fin_make_untimed_def*)+
apply *clarify*
apply (*simp add: fin_inf_append_def*)
apply (*simp add: fin_make_untimed_def*)+
apply *clarify*
apply (*simp add: fin_inf_append_def*)
done

lemma *L2_Controller*:
⟦ *Controller_L y (fin_inf_append [Zero] l) l z;*
l t = Zero ⟧
⟹ *last (fin_make_untimed (inf_truncate z t)) = Zero*
apply (*induct t*)
apply (*simp add: Controller_L_def*)
apply *clarify*
apply (*erule_tac x=0* **in** *allE*)
apply (*simp split add: split_if_asm*)
apply (*simp add: fin_make_untimed_def*)+
apply (*subgoal_tac*
  Controller_L y (fin_inf_append [Zero] l) l z)
  prefer *2*
  apply *simp*
apply (*simp add: Controller_L_def*)
apply *clarify*
apply (*erule_tac x=Suc t* **in** *allE*)
apply (*erule_tac x=t* **in** *allE*)
apply (*simp split add: split_if_asm*)
apply (*simp add: correct_fin_inf_append1*)+
done

lemma *L3_Controller*:
⟦ *Controller_L y (fin_inf_append [Zero] l) l z*⟧
⟹
last (fin_make_untimed (inf_truncate z t)) = l t
apply (*case_tac l t*)
apply (*simp add: L1_Controller L2_Controller*)+
done

lemma *L4_Controller*:
⟦ *Controller_L s (fin_inf_append [Zero] l) l z* ⟧
⟹ *fin_make_untimed (inf_truncate z i) ≠ []*
apply (*simp add: Controller_L_def*)
apply (*subgoal_tac ∀ i. 0 ≤ i ⟶ fin_make_untimed (inf_truncate z i) ≠ []*)
  prefer *2*
  apply (*simp add: fin_make_untimed_inf_truncate_Nonempty_all0a*)
apply *simp*
done

B.2.2. Properties of the System

lemma *L1_ControlSystem*:

ControlSystemArch s $\implies$ *(200::nat)* $\leq$ *hd (s i)*
  apply *(simp only: ControlSystemArch_def)*
  apply *clarify*
  apply *(induct i)*
  apply *(simp add: SteamBoiler_def)*
  apply *(simp add: Converter_def)*
  apply *atomize*
  apply *(erule_tac x=x in allE)*
  apply *(erule_tac x=z in allE)*
  apply *(erule_tac x=y in allE)*
  apply *simp*
  apply *(simp add: SteamBoiler_def Controller_def Converter_def)*
  apply *clarify*
  apply *(erule_tac x=i in allE)+*
  apply *clarify*
  apply *(simp split add: split_if_asm)*

  apply *(simp add: L4_Controller)*

  apply *(subgoal_tac last (fin_make_untimed (inf_truncate z i)) = l i)*
    prefer *2*
    apply *(simp add: L3_Controller)*
  apply *(simp add: Controller_L_def)*
  apply *clarify*
  apply *(erule_tac x=i in allE)*
  apply *(simp split add: split_if_asm)*
  apply *arith*
  apply *(simp add: fin_make_untimed_nth_length)*
  apply *(simp add: last_nth_length)*
  apply *arith*
  done

lemma *L2_ControlSystem:*
ControlSystemArch s $\implies$ *hd (s i)* $\leq$ *(800:: nat)*
  apply *(simp only: ControlSystemArch_def)*
  apply *clarify*
  apply *(induct i)*
  apply *(simp add: SteamBoiler_def)*
  apply *(simp add: Converter_def)*
  apply *atomize*
  apply *(erule_tac x=x in allE)*
  apply *(erule_tac x=z in allE)*
  apply *(erule_tac x=y in allE)*
  apply *simp*
  apply *(simp add: SteamBoiler_def Controller_def Converter_def)*
  apply *clarify*
  apply *(erule_tac x=i in allE)*
  apply *(erule_tac x=i in allE)*
  apply *clarify*
  apply *(simp split add: split_if_asm)*

  apply *(simp add: L4_Controller)*

  apply *(subgoal_tac last (fin_make_untimed (inf_truncate z i)) = l i)*
    prefer *2*
    apply *(simp add: L3_Controller)*

> **apply** (*simp add: Controller_L_def*)
> **apply** *clarify*
> **apply** (*erule_tac x=i in allE*)
> **apply** (*simp split add: split_if_asm*)
> **apply** *arith+*
>
> **apply** (*subgoal_tac last (fin_make_untimed (inf_truncate z i)) = l i*)
>   **prefer** *2*
>   **apply** (*simp add: L3_Controller*)
> **apply** (*simp add: Controller_L_def*)
> **apply** *clarify*
> **apply** (*erule_tac x=i in allE*)
> **apply** (*simp split add: split_if_asm*)
> **apply** (*simp add: last_nth_length*)+
> **done**

B.2.3. Proof of the Refinement Relation

lemma *L0_ControlSystem*:
⟦ *ControlSystemArch s*⟧ ⟹ *ControlSystem s*
>   **apply** (*simp add: ControlSystem_def*)
>   **apply** *auto*
>   **apply** (*simp add: ControlSystemArch_def*)
>   **apply** (*simp add: SteamBoiler_def*)
>   **apply** (*simp add: Converter_def*)
>   **apply** *auto*
>   **apply** (*simp add: L1_ControlSystem*)
>   **apply** (*simp add: L2_ControlSystem*)
>   **done**

B.3. Theory FR - System Specification

theory *FR = Main + FR_types*:

B.3.1. Auxiliary predicates

constdefs
>   *DisjointSchedules* :: *nat* ⟹ *nConfig* ⟹ *bool*
>   *DisjointSchedules n nC*
>
>   ≡
>   ∀ *i j. i < n ∧ j < n ∧ i ≠ j* ⟶
>   *disjoint* (*schedule* (*nC i*)) (*schedule* (*nC j*))

constdefs
>   *IdenticCycleLength* :: *nat* ⟹ *nConfig* ⟹ *bool*
>   *IdenticCycleLength n nC*
>
>   ≡
>   ∀ *i j. i < n ∧ j < n* ⟶
>   *cycleLength* (*nC i*) = *cycleLength* (*nC j*)

constdefs
>   *FrameTransmission* ::
>   *nat* ⟹ *'a nFrame* ⟹ *'a nFrame* ⟹ *nNat* ⟹ *nConfig* ⟹ *bool*
>   *FrameTransmission n nStore nReturn nGet nC*

$\equiv$
$\forall \ (t::nat) \ (k::nat). \ k < n \longrightarrow$
$(\ let \ s = t \ mod \ (cycleLength \ (nC \ k))$
$\quad in$
$\quad (\ s \ mem \ (schedule \ (nC \ k))$
$\qquad \longrightarrow$
$\qquad (nGet \ k \ t) = [s] \ \wedge$
$\qquad (\forall \ j. \ j < n \wedge j \neq k \longrightarrow$
$\qquad \quad ((nStore \ j) \ t) = ((nReturn \ k) \ t)) \ ))$

constdefs
$\quad Broadcast ::$
$\qquad nat \Rightarrow {}'a \ nFrame \Rightarrow {}'a \ Frame \ istream \Rightarrow bool$
$\quad Broadcast \ n \ nSend \ recv$
$\qquad \equiv$
$\forall \ (t::nat).$
$(\ if \ \exists \ k. \ k < n \wedge ((nSend \ k) \ t) \neq []$
$\quad then \ (recv \ t) = ((nSend \ (SOME \ k. \ k < n \wedge ((nSend \ k) \ t) \neq [])) \ t)$
$\quad else \ (recv \ t) = [] \ )$

constdefs
$\quad Receive ::$
$\quad {}'a \ Frame \ istream \Rightarrow {}'a \ Frame \ istream \Rightarrow nat \ istream \Rightarrow bool$
$\quad Receive \ recv \ store \ activation$
$\qquad \equiv$
$\forall \ (t::nat).$
$(\ if \ (activation \ t) = []$
$\quad then \ (store \ t) = (recv \ t)$
$\quad else \ (store \ t) = [])$

constdefs
$\quad Send ::$
$\quad {}'a \ Frame \ istream \Rightarrow {}'a \ Frame \ istream \Rightarrow nat \ istream \Rightarrow nat \ istream \Rightarrow bool$
$\quad Send \ return \ send \ get \ activation$
$\qquad \equiv$
$\forall \ (t::nat).$
$(\ if \ (activation \ t) = []$
$\quad then \ (get \ t) = [] \wedge (send \ t) = []$
$\quad else \ (get \ t) = (activation \ t) \wedge (send \ t) = (return \ t) \ )$

B.3.2. Main definitions

constdefs
$\quad FlexRay ::$
$\quad nat \Rightarrow {}'a \ nFrame \Rightarrow nConfig \Rightarrow {}'a \ nFrame \Rightarrow nNat \Rightarrow bool$
$\quad FlexRay \ n \ nReturn \ nC \ nStore \ nGet$
$\qquad \equiv$
$(CorrectSheaf \ n) \ \wedge$
$((\forall \ (i::nat). \ i < n \longrightarrow (msg \ 1 \ (nReturn \ i))) \ \wedge$
$(DisjointSchedules \ n \ nC) \wedge (IdenticCycleLength \ n \ nC)$
$\longrightarrow$
$(FrameTransmission \ n \ nStore \ nReturn \ nGet \ nC) \wedge$
$(\forall \ (i::nat). \ i < n \longrightarrow (msg \ 1 \ (nGet \ i)) \wedge (msg \ 1 \ (nStore \ i))) \ )$

constdefs

   Cable :: *nat* $\Rightarrow$ *'a nFrame* $\Rightarrow$ *'a Frame istream* $\Rightarrow$ *bool*
   Cable n nSend recv

   $\equiv$

   (*CorrectSheaf n*)

   $\wedge$

   ((*inf_disj n nSend*) $\longrightarrow$ (*Broadcast n nSend recv*))

constdefs

   Scheduler :: *Config* $\Rightarrow$ *nat istream* $\Rightarrow$ *bool*
   Scheduler c activation

   $\equiv$

   $\forall$ (*t::nat*).
   (*let s* = (*t mod* (*cycleLength c*))
     in
      (*if* (*s mem* (*schedule c*))
      then (*activation t*) = [*s*]
      else (*activation t*) = []))

constdefs

   BusInterface ::
    nat istream $\Rightarrow$ *'a Frame istream* $\Rightarrow$ *'a Frame istream* $\Rightarrow$
    'a Frame istream $\Rightarrow$ *'a Frame istream* $\Rightarrow$ *nat istream* $\Rightarrow$ *bool*
   BusInterface activation return recv store send get

   $\equiv$

   (*Receive recv store activation*) $\wedge$
   (*Send return send get activation*)

constdefs

   FlexRayController ::
    'a Frame istream $\Rightarrow$ *'a Frame istream* $\Rightarrow$ *Config* $\Rightarrow$
    'a Frame istream $\Rightarrow$ *'a Frame istream* $\Rightarrow$ *nat istream* $\Rightarrow$ *bool*
   FlexRayController return recv c store send get

   $\equiv$

   ($\exists$ *activation*.
    (*Scheduler c activation*) $\wedge$
    (*BusInterface activation return recv store send get*))

constdefs

   FlexRayArchitecture ::
   nat $\Rightarrow$ *'a nFrame* $\Rightarrow$ *nConfig* $\Rightarrow$ *'a nFrame* $\Rightarrow$ *nNat* $\Rightarrow$ *bool*
   FlexRayArchitecture n nReturn nC nStore nGet

   $\equiv$

   (*CorrectSheaf n*) $\wedge$
   ($\exists$ *nSend recv*.
   (*Cable n nSend recv*) $\wedge$
   ($\forall$ (*i::nat*). *i* < *n* $\longrightarrow$
     FlexRayController (*nReturn i*) *recv* (*nC i*)
        (*nStore i*) (*nSend i*) (*nGet i*)))

constdefs

   FlexRayArch ::

$nat \Rightarrow {}'a\ nFrame \Rightarrow nConfig \Rightarrow {}'a\ nFrame \Rightarrow nNat \Rightarrow bool$
FlexRayArch n nReturn nC nStore nGet

$\equiv$

$(CorrectSheaf\ n)\ \wedge$
$((\forall\ (i::nat).\ i < n \longrightarrow msg\ 1\ (nReturn\ i))\ \wedge$
$(DisjointSchedules\ n\ nC)\ \wedge\ (IdenticCycleLength\ n\ nC)$
$\longrightarrow$

$(FlexRayArchitecture\ n\ nReturn\ nC\ nStore\ nGet))$

B.4. Proof of the FlexRay System Properties

theory *FR_proof = Main + FR*:

lemma *disjointFrame_lemma*:
⟦ *DisjointSchedules n nC*; *0 < n*;
IdenticCycleLength n nC;
$\forall\ i < n.$ *FlexRayController (nReturn i) rcv*
  (nC i) (nStore i) (nSend i) (nGet i) ⟧
$\Longrightarrow$ *inf_disj n nSend*
  apply *(subgoal_tac*
    $\forall i < n.$ *FlexRayController (nReturn i) rcv*
      (nC i) (nStore i) (nSend i) (nGet i))
    prefer *2*
    apply *simp*
  apply *(simp only: inf_disj_def)*
  apply *clarify*
  apply *(erule_tac x=i in allE)*
  apply *(erule_tac x=j in allE)*
  apply *(simp add: FlexRayController_def)*
  apply *clarify*
  apply *(simp add: DisjointSchedules_def)*
  apply *(erule_tac x=i in allE)*
  apply *(erule_tac x=j in allE)*
  apply *(simp add: BusInterface_def)*
  apply *clarify*
  apply *(simp add: Send_def)*
  apply *(erule_tac x=t in allE)+*
  apply *(simp split add: split_if_asm)*
  apply *clarify*
  apply *(simp add: Scheduler_def)*
  apply *(erule_tac x=t in allE)+*
  apply *(simp add: Let_def)*
  apply *(simp split add: split_if_asm)*
  apply *(simp only: IdenticCycleLength_def)*
  apply *(erule_tac x=i in allE)*
  apply *(erule_tac x=j in allE)*
  apply *(subgoal_tac ¬ disjoint (schedule (nC i)) (schedule (nC j)))*
    prefer *2*
    apply *(erule mem_notdisjoint)*
    apply *auto*
  done

lemma *correct_DisjointSchedules1*:
⟦ *DisjointSchedules n nC*; *IdenticCycleLength n nC*;

$(t \bmod cycleLength\ (nC\ k))\ mem\ schedule\ (nC\ k);$
$k < n;\ j < n;\ k \neq j\ ]$
$\Longrightarrow$
$\neg\ (t \bmod cycleLength\ (nC\ j)\ mem\ schedule\ (nC\ j))$
apply (*simp add: DisjointSchedules_def*)
apply (*erule_tac x=k in allE*)
apply (*erule_tac x=j in allE*)
apply *clarify*
apply (*simp only: IdenticCycleLength_def*)
apply (*erule_tac x=k in allE*)
apply (*erule_tac x=j in allE*)
apply *simp*
apply (*subgoal_tac ¬ disjoint (schedule (nC k)) (schedule (nC j))*)
  prefer *2*
  apply (*erule mem_notdisjoint*)
  apply *auto*
done

lemma *fr_Send*:
[*FlexRayController* (*nReturn i*) *recv* (*nC i*)
  (*nStore i*) (*nSend i*) (*nGet i*);
  $\neg\ (t \bmod cycleLength\ (nC\ i)\ mem\ schedule\ (nC\ i))$]
  $\Longrightarrow$
  (*nSend i*) $t = []$
apply (*simp add: FlexRayController_def*)
apply (*simp add: Scheduler_def*)
apply *clarify*
apply (*erule_tac x=t in allE*)
apply (*simp add: Let_def*)
apply (*simp add: BusInterface_def*)
apply *clarify*
apply (*simp add: Send_def*)
done

lemma *fr_nC_Send*:
[$\forall i < n.$ *FlexRayController* (*nReturn i*) *recv*
     (*nC i*) (*nStore i*) (*nSend i*) (*nGet i*);
  $0 < n;\ k < n;$
  DisjointSchedules n nC;
  IdenticCycleLength n nC;
  $t \bmod cycleLength\ (nC\ k)\ mem\ schedule\ (nC\ k)$]
  $\Longrightarrow$
  $\forall j.\ j < n \wedge j \neq k \longrightarrow (nSend\ j)\ t = []$
apply *clarify*
apply (*subgoal_tac*
  $\neg\ (t \bmod cycleLength\ (nC\ j)\ mem\ schedule\ (nC\ j))$)
  prefer *2*
  apply (*erule correct_DisjointSchedules1*)
  apply *assumption+*
  apply *simp*
apply (*erule_tac x=j in allE*)
apply (*simp add: fr_Send*)
done

lemma *fr_nStore_nReturn*:
⟦*Cable n nSend recv*;
  ∀ *i<n. FlexRayController* (*nReturn i*) *recv*
      (*nC i*) (*nStore i*) (*nSend i*) (*nGet i*);
  0 < n; *k < n*;
  DisjointSchedules n nC;
  IdenticCycleLength n nC;
  t mod cycleLength (*nC k*) *mem schedule* (*nC k*)⟧
  ⟹
  ∀ *j. j < n ∧ j ≠ k ⟶ nStore j t = nReturn k t*
apply *clarify*
apply (*subgoal_tac*
  ∀ *j. j < n ∧ j ≠ k ⟶* (*nSend j*) *t* = [])
  prefer *2*
  apply (*simp add: fr_nC_Send*)
apply (*simp add: Cable_def*)
apply (*simp add: CorrectSheaf_def*)
apply (*subgoal_tac inf_disj n nSend*)
  prefer *2*
  apply (*simp add: disjointFrame_lemma*)
apply *simp*
apply (*subgoal_tac*
  ∀ *i<n. FlexRayController* (*nReturn i*) *recv*
      (*nC i*) (*nStore i*) (*nSend i*) (*nGet i*))
  prefer *2*
  apply *simp*
apply (*erule_tac x=j* **in** *allE*)
apply (*rotate_tac −2*)
apply (*erule_tac x=k* **in** *allE*)
apply (*simp add: Broadcast_def*)
apply (*erule_tac x=t* **in** *allE*)
apply (*simp add: FlexRayController_def*)
apply (*simp add: Scheduler_def*)
apply *clarify*
apply (*rotate_tac −4*)
apply (*erule_tac x=t* **in** *allE*)
apply (*erule_tac x=t* **in** *allE*)
apply (*simp add: Let_def*)

apply (*case_tac* (*t mod cycleLength* (*nC j*) *mem schedule* (*nC j*)))
apply (*simp add: correct_DisjointSchedules1*)

apply (*simp add: BusInterface_def*)
apply *clarify*
apply (*simp add: Send_def*)
apply (*rotate_tac −3*)
apply (*erule_tac x=t* **in** *allE*)
apply (*erule_tac x=t* **in** *allE*)
apply (*simp split add: split_if_asm*)
apply (*subgoal_tac*
  (*SOME i. i < n ∧ nSend i t ≠* []) = *ka*)
  prefer *2*
  apply (*simp add: inf_disj_index*)
apply *simp*
apply (*erule_tac*
  V=(*SOME i. i < n ∧ nSend i t ≠* []) = *ka*
  in *thin_rl*)

apply (*simp only*: *inf_disj_def*)
apply (*erule_tac* *x=ka* **in** *allE*)
apply (*erule_tac* *x=t* **in** *allE*)
apply (*erule_tac* *x=ka* **in** *allE*)
apply (*erule_tac* *x=k* **in** *allE*)
apply (*simp add*: *Receive_def*)+
done

lemma *fr_refinement_FrameTransmission*:
⟦ *Cable n nSend recv*;
  ∀ *i*<*n*. *FlexRayController* (*nReturn i*) *recv*
      (*nC i*) (*nStore i*) (*nSend i*) (*nGet i*);
  0 < *n*;
  DisjointSchedules n nC; *IdenticCycleLength n nC*⟧
⟹ *FrameTransmission n nStore nReturn nGet nC*
apply (*simp add*: *FrameTransmission_def*)
apply *clarify*
apply (*simp add*: *Let_def*)
apply *clarify*
apply (*simp add*: *fr_nStore_nReturn*)
apply (*erule_tac* *x=k* **in** *allE*)
apply (*simp add*: *FlexRayController_def*)
apply (*simp add*: *BusInterface_def*)
apply *clarify*
apply (*simp add*: *Send_def Scheduler_def*)
apply (*erule_tac* *x=t* **in** *allE*)+
apply (*simp add*: *Let_def*)
done

lemma *fr_refinement_msg_nGet*:
⟦ *i* < *n*;
  ∀ *i*<*n*. *FlexRayController* (*nReturn i*) *recv*
      (*nC i*) (*nStore i*) (*nSend i*) (*nGet i*)⟧
⟹ *msg* (*Suc 0*) (*nGet i*)
apply (*simp add*: *FlexRayController_def*)
apply (*erule_tac* *x=i* **in** *allE*)
apply *clarify*
apply (*simp add*: *BusInterface_def msg_def*)
apply (*simp add*: *Send_def Scheduler_def*)
apply *clarify*
apply (*erule_tac* *x=t* **in** *allE*)+
apply (*simp add*: *Let_def*)
apply (*simp split add*: *split_if_asm*)
done

lemma *fr_refinement_msg_nSend*:
⟦ *msg* (*Suc 0*) (*nReturn i*);
  BusInterface activation (*nReturn i*) *recv*
    (*nStore i*) (*nSend i*) (*nGet i*)⟧
⟹ *msg* (*Suc 0*) (*nSend i*)
apply (*simp add*: *msg_def BusInterface_def*)
apply *clarify*
apply (*simp add*: *Send_def*)
apply (*erule_tac* *x=t* **in** *allE*)+

apply (*simp split add: split_if_asm*)
done

lemma *fr_refinement_msg_nStore*:
⟦ *DisjointSchedules n nC*; *IdenticCycleLength n nC*;
 inf_disj n nSend; *i < n*; *0 < n*;
 ∀ *i<n. msg* (*Suc 0*) (*nReturn i*);
 Cable n nSend recv;
 ∀ *i<n. FlexRayController* (*nReturn i*) *recv*
        (*nC i*) (*nStore i*) (*nSend i*) (*nGet i*)⟧
⟹ *msg* (*Suc 0*) (*nStore i*)
 apply (*simp* (*no_asm*) *add: msg_def*)
 apply (*simp add: Cable_def*)
 apply *clarify*
 apply (*simp add: CorrectSheaf_def Broadcast_def*)
 apply (*rotate_tac −1*)
 apply (*erule_tac x=t* **in** *allE*)
 apply (*simp split add: split_if_asm*)
 apply (*subgoal_tac*
   (*SOME i. i < n* ∧ *nSend i t ≠* []) = *k*)
  prefer *2*
  apply (*simp add: inf_disj_index*)
 apply *simp*
 apply (*erule_tac*
   V=(*SOME i. i < n* ∧ *nSend i t ≠* []) = *k*
   in *thin_rl*)
 apply (*simp only: inf_disj_def*)
 apply (*rotate_tac 2*)
 apply (*erule_tac x=t* **in** *allE*)
 apply (*erule_tac x=k* **in** *allE*)
 apply *simp*
 apply (*subgoal_tac*
   ∀ *i<n. FlexRayController* (*nReturn i*) *recv*
         (*nC i*) (*nStore i*) (*nSend i*) (*nGet i*))
  prefer *2*
  apply *simp*
 apply (*erule_tac x=k* **in** *allE*)
 apply (*rotate_tac −2*)
 apply (*erule_tac x=i* **in** *allE*)
 apply *clarify*
 apply (*simp add: FlexRayController_def*)
 apply (*erule_tac x=k* **in** *allE*)
 apply *clarify*
 apply (*subgoal_tac msg* (*Suc 0*) (*nSend k*))
  prefer *2*
  apply (*erule fr_refinement_msg_nSend*)
  apply *assumption*
 apply (*simp add: BusInterface_def*)
 apply *clarify*
 apply (*simp add: Receive_def*)
 apply (*rotate_tac −4*)
 apply (*erule_tac x=t* **in** *allE*)
 apply (*erule_tac x=t* **in** *allE*)
 apply (*simp split add: split_if_asm*)
 apply (*simp add: msg_def*)+
 apply (*rotate_tac 5*)

```
    apply (erule_tac x=i in allE)
    apply (simp add: FlexRayController_def)
    apply (simp add: BusInterface_def)
    apply clarify
    apply (simp add: Receive_def)
    apply (rotate_tac −2)
    apply (erule_tac x=t in allE)
    apply (simp split add: split_if_asm)
    done


lemma fr_refinement_msg:
⟦ Cable n nSend recv;
    i < n;   0 < n;   ∀ i<n. msg (Suc 0) (nReturn i);
    DisjointSchedules n nC;  IdenticCycleLength n nC;
    ∀ i<n. FlexRayController (nReturn i) recv
             (nC i) (nStore i) (nSend i) (nGet i)⟧
⟹ msg (Suc 0) (nGet i) ∧ msg (Suc 0) (nStore i)
    apply (subgoal_tac inf_disj n nSend)
      prefer 2
      apply (simp add: disjointFrame_lemma)
    apply (rule conjI)
    apply (simp add: fr_refinement_msg_nGet)
    apply (erule fr_refinement_msg_nStore)
    apply auto
    done


theorem main_fr_refinement:
FlexRayArch n nReturn nC nStore nGet
⟹ FlexRay n nReturn nC nStore nGet
    apply (simp add: FlexRayArch_def FlexRay_def)
    apply (simp add: FlexRayArchitecture_def)
    apply (simp add: CorrectSheaf_def)
    apply auto
    apply (simp add: fr_refinement_FrameTransmission)
    apply (simp only: fr_refinement_msg)+
    done


end
```

B.5. Automotive-Gateway System Specification

theory *Gateway* = *Main* + *Gateway_types*:

constdefs
ServiceCenter ::
    ECall_Info istream ⟹ *aType istream* ⟹ *bool*
ServiceCenter i a
    ≡
    ∀ (t::nat).
    a 0 = [] ∧   *a (Suc t)* =   (*if (i t)* = [] *then* [] *else* [sc_ack])

constdefs
  Loss ::
    bool istream ⇒ aType istream ⇒ ECall_Info istream ⇒
    aType istream ⇒ ECall_Info istream ⇒ bool
  Loss lose a i2 a2 i
  ≡
  ∀ (*t::nat*).
  (*if lose t* = [*False*]
    then a2 t = *a t* ∧ *i t* = *i2 t*
    else a2 t = [] ∧ *i t* = [])

constdefs
  Delay ::
    aType istream ⇒ ECall_Info istream ⇒ nat ⇒
    aType istream ⇒ ECall_Info istream ⇒ bool
  Delay a2 i1 d a1 i2
  ≡
  ∀ (*t::nat*).
  ($t < d$ ⟶ *a1 t* = [] ∧ *i2 t* = []) ∧
  (*a1* (*t+d*) = *a2 t*) ∧
  (*i2* (*t+d*) = *i1 t*)

constdefs
  tiTable_SampleT ::
  reqType istream ⇒ aType istream ⇒
  stopType istream ⇒ bool istream ⇒
  (*nat ⇒ GatewayStatus*) ⇒ (*nat ⇒ ECall_Info list*) ⇒
  GatewayStatus istream ⇒ ECall_Info istream ⇒ vcType istream
  ⇒ (*nat ⇒ GatewayStatus*) ⇒ *bool*
  tiTable_SampleT req a1 stop lose st_in buffer_in
      ack i1 vc st_out
  ≡
  ∀ (*t::nat*)
    (*r::reqType list*) (*x::aType list*)
    (*y::stopType list*) (*z::bool list*).
  (*∗1∗*)
  (*st_in t* = *init_state* ∧ *req t* = [*init*]
    ⟶ *ack t* = [*call*] ∧ *i1 t* = [] ∧ *vc t* = []
    ∧ *st_out t* = *call*)
  ∧
  (*∗2∗*)
  (*st_in t* = *init_state* ∧ *req t* ≠ [*init*]
    ⟶ *ack t* = [*init_state*] ∧ *i1 t* = [] ∧ *vc t* = []
    ∧ *st_out t* = *init_state*)
  ∧
  (*∗3∗*)
  ((*st_in t* = *call* ∨ (*st_in t* = *connection_ok* ∧ *r* ≠ [*send*])) ∧
  req t = *r* ∧ *lose t* = [*False*]
    ⟶ *ack t* = [*connection_ok*] ∧ *i1 t* = [] ∧ *vc t* = []
    ∧ *st_out t* = *connection_ok*)
  ∧
  (*∗4∗*)
  ((*st_in t* = *call* ∨ *st_in t* = *connection_ok* ∨ *st_in t* = *sending_data*)
    ∧ *lose t* = [*True*]
    ⟶ *ack t* = [*init_state*] ∧ *i1 t* = [] ∧ *vc t* = []

$$\wedge\ st\_out\ t = init\_state\ )$$
$$\wedge$$
$(*5*)$
$$(\ st\_in\ t = connection\_ok \wedge req\ t = [send] \wedge lose\ t = [False]$$
$$\longrightarrow ack\ t = [sending\_data] \wedge i1\ t = buffer\_in\ t \wedge vc\ t = []$$
$$\wedge\ st\_out\ t = sending\_data\ )$$
$$\wedge$$
$(*6*)$
$$(\ st\_in\ t = sending\_data \wedge a1\ t = [] \wedge lose\ t = [False]$$
$$\longrightarrow ack\ t = [sending\_data] \wedge i1\ t = [] \wedge vc\ t = []$$
$$\wedge\ st\_out\ t = sending\_data\ )$$
$$\wedge$$
$(*7*)$
$$(\ st\_in\ t = sending\_data \wedge a1\ t = [sc\_ack] \wedge lose\ t = [False]$$
$$\longrightarrow ack\ t = [voice\_com] \wedge i1\ t = [] \wedge vc\ t = [vc\_com]$$
$$\wedge\ st\_out\ t = voice\_com\ )$$
$$\wedge$$
$(*8*)$
$$(\ st\_in\ t = voice\_com \wedge stop\ t = [] \wedge lose\ t = [False]$$
$$\longrightarrow ack\ t = [voice\_com] \wedge i1\ t = [] \wedge vc\ t = [vc\_com]$$
$$\wedge\ st\_out\ t = voice\_com\ )$$
$$\wedge$$
$(*9*)$
$$(\ st\_in\ t = voice\_com \wedge stop\ t = [] \wedge lose\ t = [True]$$
$$\longrightarrow ack\ t = [voice\_com] \wedge i1\ t = [] \wedge vc\ t = []$$
$$\wedge\ st\_out\ t = voice\_com\ )$$
$$\wedge$$
$(*10*)$
$$(\ st\_in\ t = voice\_com \wedge stop\ t = [stop\_vc]$$
$$\longrightarrow ack\ t = [init\_state] \wedge i1\ t = [] \wedge vc\ t = []$$
$$\wedge\ st\_out\ t = init\_state\ )$$

constdefs
$Sample\_L ::$
$reqType\ istream \Rightarrow ECall\_Info\ istream \Rightarrow aType\ istream \Rightarrow$
$stopType\ istream \Rightarrow bool\ istream \Rightarrow$
$(nat \Rightarrow GatewayStatus) \Rightarrow (nat \Rightarrow ECall\_Info\ list) \Rightarrow$
$GatewayStatus\ istream \Rightarrow ECall\_Info\ istream \Rightarrow vcType\ istream$
$\Rightarrow (nat \Rightarrow GatewayStatus) \Rightarrow (nat \Rightarrow ECall\_Info\ list)$
$\Rightarrow bool$
$Sample\_L\ req\ dt\ a1\ stop\ lose\ st\_in\ buffer\_in$
$\qquad ack\ i1\ vc\ st\_out\ buffer\_out$
$\qquad \equiv$
$(\forall\ (t::nat).$
$\ buffer\_out\ t =$
$\ (if\ dt\ t = []\ then\ buffer\_in\ t\ \ else\ dt\ t)\ )$
$\wedge$
$(tiTable\_SampleT\ req\ a1\ stop\ lose\ st\_in\ buffer\_in$
$\qquad\qquad ack\ i1\ vc\ st\_out)$

constdefs
$Sample ::$
$reqType\ istream \Rightarrow ECall\_Info\ istream \Rightarrow aType\ istream \Rightarrow$
$stopType\ istream \Rightarrow bool\ istream \Rightarrow$
$GatewayStatus\ istream \Rightarrow ECall\_Info\ istream \Rightarrow vcType\ istream$
$\Rightarrow bool$

Sample req dt a1 stop lose ack i1 vc
≡
$((msg\ (1::nat)\ req) \land$
$(msg\ (1::nat)\ a1)\ \land$
$(msg\ (1::nat)\ stop))$
$\longrightarrow$
$(\exists\ st\ buffer.$
$(Sample\_L\ req\ dt\ a1\ stop\ lose$
$(fin\_inf\_append\ [init\_state]\ st)$
$(fin\_inf\_append\ [[]]\ buffer)$
$ack\ i1\ vc\ st\ buffer)\ )$

constdefs
Gateway ::
reqType istream ⇒ *ECall_Info istream* ⇒ *aType istream* ⇒
stopType istream ⇒ *bool istream* ⇒ *nat* ⇒
GatewayStatus istream ⇒ *ECall_Info istream* ⇒ *vcType istream*
⇒ *bool*
Gateway req dt a stop lose d ack i vc
≡ $\exists\ i1\ i2\ a1\ a2.$
$(Sample\ req\ dt\ a1\ stop\ lose\ ack\ i1\ vc)\ \land$
$(Delay\ a2\ i1\ d\ a1\ i2)\ \land$
$(Loss\ lose\ a\ i2\ a2\ i)$

constdefs
tiTable_SampleT_ext ::
reqType istream ⇒ *aType istream* ⇒
stopType istream ⇒ *bool istream* ⇒
$(nat ⇒ GatewayStatus) ⇒ (nat ⇒ ECall\_Info\ list) ⇒$
GatewayStatus istream ⇒ *ECall_Info istream* ⇒ *vcType istream*
⇒ $(nat ⇒ GatewayStatus) ⇒ bool$
tiTable_SampleT_ext req a1 stop lose st_in buffer_in
ack i1 vc st_out
≡
$\forall\ (t::nat)$
$(r::reqType\ list)\ (x::aType\ list)$
$(y::stopType\ list)\ (z::bool\ list).$
$(*1*)$
$(\ st\_in\ t = init\_state \land req\ t = [init]$
$\land\ a1\ t = x\ \land stop\ t = y\ \land lose\ t = z$
$\longrightarrow\ ack\ t = [call]\ \land i1\ t = []\ \land vc\ t = []$
$\land\ st\_out\ t = call\ )$
$\land$
$(*2*)$
$(\ st\_in\ t = init\_state \land r \neq [init]$
$\land\ req\ t = r\ \land a1\ t = x\ \land stop\ t = y\ \land lose\ t = z$
$\longrightarrow\ \ ack\ t = [init\_state]\ \land i1\ t = []\ \land vc\ t = []$
$\land\ st\_out\ t = init\_state\ )$
$\land$
$(*3*)$
$(\ (st\_in\ t = call \lor (st\_in\ t = connection\_ok \land r \neq [send]))\ \land$
$req\ t = r\ \land a1\ t = x\ \land stop\ t = y\ \land lose\ t = [False]$
$\longrightarrow\ ack\ t = [connection\_ok]\ \land i1\ t = []\ \land vc\ t = []$
$\land\ st\_out\ t = connection\_ok\ )$
$\land$
$(*4*)$

$(\ (st\_in \ t = call \ \lor \ st\_in \ t = connection\_ok \ \lor \ st\_in \ t = sending\_data)$
$\quad \land \ req \ t = r \ \land \ a1 \ t = x \ \land \ stop \ t = y \ \land \ lose \ t = [True]$
$\quad \longrightarrow \ ack \ t = [init\_state] \ \land \ i1 \ t = [] \ \land \ vc \ t = []$
$\qquad \land \ st\_out \ t = init\_state \ )$
$\land$
$(*5*)$
$(\ st\_in \ t = connection\_ok \ \land \ req \ t = [send] \ \land \ req \ t = r$
$\quad \land \ a1 \ t = x \ \land \ stop \ t = y \ \land \ lose \ t = [False]$
$\quad \longrightarrow \ ack \ t = [sending\_data] \ \land \ i1 \ t = buffer\_in \ t \ \land \ vc \ t = []$
$\qquad \land \ st\_out \ t = sending\_data \ )$
$\land$
$(*6*)$
$(\ st\_in \ t = sending\_data \ \land \ req \ t = r \ \land \ a1 \ t = []$
$\quad \land \ stop \ t = y \ \land \ lose \ t = [False]$
$\quad \longrightarrow \ ack \ t = [sending\_data] \ \land \ i1 \ t = [] \ \land \ vc \ t = []$
$\qquad \land \ st\_out \ t = sending\_data \ )$
$\land$
$(*7*)$
$(\ st\_in \ t = sending\_data \ \land \ req \ t = r \ \land \ a1 \ t = [sc\_ack]$
$\quad \land \ stop \ t = y \ \land \ lose \ t = [False]$
$\quad \longrightarrow \ ack \ t = [voice\_com] \ \land \ i1 \ t = [] \ \land \ vc \ t = [vc\_com]$
$\qquad \land \ st\_out \ t = voice\_com \ )$
$\land$
$(*8*)$
$(\ st\_in \ t = voice\_com \ \land \ req \ t = r \ \land \ a1 \ t = x$
$\quad \land \ stop \ t = [] \ \land \ lose \ t = [False]$
$\quad \longrightarrow \ ack \ t = [voice\_com] \ \land \ i1 \ t = [] \ \land \ vc \ t = [vc\_com]$
$\qquad \land \ st\_out \ t = voice\_com \ )$
$\land$
$(*9*)$
$(\ st\_in \ t = voice\_com \ \land \ req \ t = r \ \land \ a1 \ t = x$
$\quad \land \ stop \ t = [] \ \land \ lose \ t = [True]$
$\quad \longrightarrow \ ack \ t = [voice\_com] \ \land \ i1 \ t = [] \ \land \ vc \ t = []$
$\qquad \land \ st\_out \ t = voice\_com \ )$
$\land$
$(*10*)$
$(\ st\_in \ t = voice\_com \ \land \ req \ t = r \ \land \ a1 \ t = x$
$\quad \land \ stop \ t = [stop\_vc] \ \land \ lose \ t = z$
$\quad \longrightarrow \ ack \ t = [init\_state] \ \land \ i1 \ t = [] \ \land \ vc \ t = []$
$\qquad \land \ st\_out \ t = init\_state \ )$

constdefs

$GatewaySystem \ ::$
$reqType \ istream \ \Rightarrow \ ECall\_Info \ istream \ \Rightarrow$
$stopType \ istream \ \Rightarrow \ bool \ istream \ \Rightarrow \ nat \ \Rightarrow$
$GatewayStatus \ istream \ \Rightarrow \ vcType \ istream$
$\Rightarrow \ bool$

$GatewaySystem \ req \ dt \ stop \ lose \ d \ ack \ vc$
$\equiv$
$\exists \ a \ i.$
$(Gateway \ req \ dt \ a \ stop \ lose \ d \ ack \ i \ vc) \ \land$
$(ServiceCenter \ i \ a)$

constdefs

GatewayReq ::
  reqType istream ⇒ *ECall_Info istream* ⇒ *aType istream* ⇒
  stopType istream ⇒ *bool istream* ⇒ *nat* ⇒
  GatewayStatus istream ⇒ *ECall_Info istream* ⇒ *vcType istream*
  ⇒ *bool*

GatewayReq req dt a stop lose d ack i vc

≡

$((msg\ (1::nat)\ req) \wedge\ (msg\ (1::nat)\ a)\ \wedge$
$(msg\ (1::nat)\ stop) \wedge\ (ts\ lose))$
⟶
$(\forall\ (t::nat).$
$(\ ack\ t = [init\_state] \wedge req\ (Suc\ t) = [init]\ \wedge$
$lose\ (t+1) = [False] \wedge lose\ (t+2) = [False]$
$\longrightarrow ack\ (t+2) = [connection\_ok])$
$\wedge$
$(\ ack\ t = [connection\_ok] \wedge req\ (Suc\ t)\ = [send]\ \wedge$
$(\forall\ (k::nat).\ k \leq (d+1) \longrightarrow lose\ (t+k) = [False])$
$\longrightarrow i\ ((Suc\ t) + d) = inf\_last\_ti\ dt\ t$
$\quad \wedge\ \ ack\ (Suc\ t) = [sending\_data])$
$\wedge$
$(\ ack\ (t+d) = [sending\_data] \wedge a\ (Suc\ t) = [sc\_ack]\ \wedge$
$(\forall\ (k::nat).\ k \leq (d+1) \longrightarrow lose\ (t+k) = [False])$
$\longrightarrow vc\ ((Suc\ t) + d) = [vc\_com])\ )$

constdefs
  GatewaySystemReq ::
  reqType istream ⇒ *ECall_Info istream* ⇒
  stopType istream ⇒ *bool istream* ⇒ *nat* ⇒
  GatewayStatus istream ⇒ *vcType istream*
  ⇒ *bool*

GatewaySystemReq req dt stop lose d ack vc

≡

$((msg\ (1::nat)\ req) \wedge (msg\ (1::nat)\ stop) \wedge (ts\ lose))$
⟶
$(\forall\ (t::nat)\ (k::nat).$
$(\ ack\ t = [init\_state] \wedge req\ (Suc\ t) = [init]$
$\wedge\ (\forall\ t1.\ t1 \leq t \longrightarrow req\ t1 = [])$
$\wedge\ req\ (t+2) = []$
$\wedge\ (\forall\ m.\ m < k + 3 \longrightarrow req\ (t + m) \neq [send])$
$\wedge\ req\ (t+3+k)\ = [send] \wedge inf\_last\_ti\ dt\ (t+2) \neq []$
$\wedge\ (\forall\ (j::nat).$
$\quad j \leq (4 + k + d + d) \longrightarrow lose\ (t+j) = [False])$
$\longrightarrow vc\ (t + 4 + k + d + d) = [vc\_com])\ )$

constdefs
  GatewayReqExt ::
  reqType istream ⇒ *ECall_Info istream* ⇒ *aType istream* ⇒
  stopType istream ⇒ *bool istream* ⇒ *nat* ⇒
  GatewayStatus istream ⇒ *ECall_Info istream* ⇒ *vcType istream*
  ⇒ *bool*

GatewayReqExt req dt a stop lose d ack i vc

≡

$((msg\ (1::nat)\ req) \wedge\ (msg\ (1::nat)\ a)\ \wedge$
$(msg\ (1::nat)\ stop) \wedge\ (ts\ lose))$
$\longrightarrow$
$(\forall\ (t::nat)\ (k::nat).$
$(\ ack\ t = [init\_state] \wedge req\ (Suc\ t) = [init] \wedge$
$\quad lose\ (t+1) = [False] \wedge lose\ (t+2) = [False]$
$\quad \longrightarrow ack\ (t+2) = [connection\_ok])$
$\wedge$
$(\ ack\ t = [init\_state] \wedge req\ (Suc\ t) = [init] \wedge$
$\quad req\ (t + 3 + k) = [send] \wedge$
$\quad (\forall\ t1 \leq t.\ req\ t1 = []) \wedge$
$\quad (\forall\ m < (k + 3).\ req\ (t + m) \neq [send]) \wedge$
$\quad (\forall\ j \leq (k + d + 3).\ lose\ (t+j) = [False])$
$\quad \longrightarrow (\forall\ t2 < (t + 3 + k + d).\ i\ t2 = [])\ )$
$\wedge$
$(\ ack\ t = [connection\_ok] \wedge$
$\quad (\forall\ m \leq k.\ req\ (t + m) \neq [send]) \wedge$
$\quad (\forall\ j \leq k.\ lose\ (t+j) = [False])$
$\quad \longrightarrow (\forall\ y \leq k.\ ack\ (t + y) = [connection\_ok])\ )$
$\wedge$
$(\ ack\ t = [connection\_ok] \wedge req\ (Suc\ t)\ = [send] \wedge$
$\quad (\forall\ (k::nat).\ k \leq (d+1) \longrightarrow lose\ (t+k) = [False])$
$\quad \longrightarrow i\ ((Suc\ t) + d) = inf\_last\_ti\ dt\ t$
$\quad\quad \wedge\ ack\ (Suc\ t) = [sending\_data])$
$\wedge$
$(\ ack\ t = [sending\_data] \wedge (\forall\ t3 \leq t+d.\ a\ t3 = []) \wedge$
$\quad (\forall\ j \leq (d+d).\ lose\ (t+j) = [False])$
$\quad \longrightarrow (\forall\ x \leq (d+d).\ ack\ (t + x) = [sending\_data]\ )\ )$
$\wedge$
$(\ ack\ (t+d) = [sending\_data] \wedge a\ (Suc\ t) = [sc\_ack] \wedge$
$\quad (\forall\ (k::nat).\ k \leq (d+1) \longrightarrow lose\ (t+k) = [False])$
$\quad \longrightarrow vc\ ((Suc\ t) + d) = [vc\_com])\ )$
end

B.6. Auxiliary Proofs for the Automotive-Gateway System

theory *Gateway_proof_aux = Main + Gateway + ts_bool_stream:*

B.6.1. Properties of the defined datatypes

lemma *aType_empty*:
$[\![msg\ (Suc\ 0)\ a;\ a\ t \neq [sc\_ack]\ ]\!] \Longrightarrow a\ t = []$
   apply (*simp add: msg_def*)
   apply (*erule_tac x=t in allE*)
   apply (*case_tac a t*)
   apply *auto*
   apply (*case_tac aa*)
   apply *auto*
   done

lemma *stopType_empty*:
$[\![msg\ (Suc\ 0)\ a;\ a\ t \neq [stop\_vc]\ ]\!] \Longrightarrow a\ t = []$
   apply (*simp add: msg_def*)
   apply (*erule_tac x=t in allE*)
   apply (*case_tac a t*)

apply *auto*
apply (*case_tac aa*)
apply *auto*
done

lemma *vcType_empty*:
⟦ *msg (Suc 0) a; a t ≠ [vc_com]* ⟧ ⟹ *a t = []*
apply (*simp add: msg_def*)
apply (*erule_tac x=t* **in** *allE*)
apply (*case_tac a t*)
apply *auto*
apply (*case_tac aa*)
apply *auto*
done

B.6.2. Equivalence of the titable representations: SampleT and SampleT_ext

lemma *univ_tiTable_Sample*:
tiTable_SampleT req a1 stop lose st_in buffer_in
$\qquad$ *ack i1 vc st_out*
=
tiTable_SampleT_ext req a1 stop lose st_in buffer_in
$\qquad$ *ack i1 vc st_out*
apply *auto*
apply (*simp add: tiTable_SampleT_ext_def tiTable_SampleT_def*)
apply (*simp add: tiTable_SampleT_def*)
apply *clarify*
apply (*simp add: tiTable_SampleT_ext_def*)
apply (*erule_tac x=t* **in** *allE*)
apply (*erule_tac x=req t* **in** *allE*)
apply (*erule_tac x=a1 t* **in** *allE*)
apply (*erule_tac x=stop t* **in** *allE*)
apply (*erule_tac x=lose t* **in** *allE*)
apply *simp*
done

B.6.3. Auxiliary Lemmas

lemma *inf_last_ti2*:
inf_last_ti dt (Suc (Suc t)) ≠ []
⟹ *inf_last_ti dt (Suc (Suc (t + k))) ≠ []*
by (*induct k, auto*)

lemma *buffer_inf_last_ti*:
⟦∀ *t. buffer t =*
$\quad$ (*if dt t = [] then fin_inf_append [[]] buffer t else dt t*) ⟧
$\quad$ ⟹
$\quad$ *buffer t = inf_last_ti dt t*
apply (*subgoal_tac*
∀ *t. buffer t =*
(*if dt t = [] then fin_inf_append [[]] buffer t else dt t*))
$\quad$ **prefer** *2*
$\quad$ **apply** *simp*
apply (*erule_tac x=t* **in** *allE*)

apply (*simp split add: split_if_asm*)
apply (*induct t*)
apply (*simp add: fin_inf_append_def*)
apply *simp*
apply (*simp add: correct_fin_inf_append1*)
apply *clarify*
apply (*case_tac dt t = []*)
apply (*simp add: correct_fin_inf_append1*)
apply (*erule_tac x=t in allE*)
apply *simp*
done

lemma *aux_ack_t2*:
$[\![\,\forall m \leq k.\ ack\ (Suc\ (Suc\ (t\ +\ m))) = [connection\_ok];$
$Suc\ (Suc\ t) < t2;\ t2 < t + 3 + k]\!]$
$\implies ack\ t2 = [connection\_ok]$
apply (*erule_tac x=(t2 − t − (2::nat)) in allE*)
apply *simp*
apply (*subgoal_tac t2 − Suc (Suc t) ≤ k*)
  prefer *2*
  apply *arith*
apply (*subgoal_tac (Suc (Suc (t2 − 2))) = t2*)
  prefer *2*
  apply *arith*
apply *simp*
done

lemma *aux_lemma_lose_1*:
$\forall j \leq (2::nat) * d + ((4::nat) + k).\ lose\ (t + j) = x \implies$
$\forall ka \leq Suc\ d.\ lose\ (Suc\ (Suc\ (t + k + ka))) = x$
apply *auto*
apply (*erule_tac x=k + 2 + ka in allE*)
apply *simp*
apply (*subgoal_tac (t + (k + ka))= (t + k + ka))*)
  prefer *2*
  apply *simp*
apply *simp*
done

lemma *aux_lemma_lose_2*:
$\forall j \leq (2::nat) * d + ((4::nat) + k).\ lose\ (t + j) = [False]$
$\implies \forall x \leq d + (1::nat).\ lose\ (t + x) = [False]$
  by *auto*

lemma *aux_lemma_lose_3*:
$\forall j \leq 2 * d + (4 + k).\ lose\ (t + j) = [False]$
$\implies \forall ka \leq Suc\ d.\ lose\ (d + (t + (3 + k)) + ka) = [False]$
apply *auto*
apply (*erule_tac x=(d + 3 + k + ka) in allE*)
apply *simp*
apply (*subgoal_tac*
  (t + (d + 3 + k + ka)) = (d + (t + (3 + k)) + ka))
  prefer *2*
  apply *arith*
apply *simp*

 done

lemma *aux_arith1_Gateway7*:
$[\![$ *t2* − *t* ≤ *(2::nat)* ∗ *d* + *(t* + *((4::nat)* + *k))*;
 t2 < *t* + *(3::nat)* + *k* + *d*; ¬ *t2* − *d* < *(0::nat)* $]\!]$
$\implies$ *t2* − *d* < *t* + *(3::nat)* + *k*
 by *arith*

lemma *Loss_Delay_msg_a*:
$[\![$ *msg (Suc 0) a*; *Delay a2 i1 d a1 i2*; *Loss lose a i2 a2 i* $]\!]$
$\implies$ *msg (Suc 0) a1*
 apply *(simp add: msg_def)*
 apply *clarify*
 apply *(simp add: Delay_def Loss_def)*
 apply *(case_tac t < d)*
 apply *(erule_tac x=t in allE)+*
 apply *simp*
 apply *(rotate_tac 1)*
 apply *(erule_tac x=t−d in allE)*
 apply *simp*
 apply *(erule_tac x=t−d in allE)*
 apply *(simp split add: split_if_asm)*
 done

lemma *tiTable_ack_st*:
$[\![$ *tiTable_SampleT req a1 stop lose st_in b ack i1 vc st_out*;
 ts lose; *msg (Suc 0) a1*; *msg (Suc 0) stop* $]\!]$
$\implies$ *ack t = [st_out t]*
 apply *(simp add: tiTable_SampleT_def)*
 apply *(erule_tac x=t in allE)*
 apply *clarify*
 apply *(case_tac st_in t)*
 apply *simp*
 apply *(case_tac req t = [init])*
 apply *simp+*
 apply *(case_tac lose t = [False])*
 apply *simp*
 apply *(subgoal_tac lose t = [True])*
  prefer *2*
  apply *(erule ts_bool_True, assumption)*
 apply *simp+*
 apply *(case_tac req t = [send])*
 apply *simp+*
 apply *(case_tac lose t = [False])*
 apply *simp*
 apply *(subgoal_tac lose t = [True])*
  prefer *2*
  apply *(erule ts_bool_True, assumption)*
 apply *simp+*
 apply *(case_tac lose t = [False])*
 apply *simp*
 apply *(subgoal_tac lose t = [True])*
  prefer *2*
  apply *(erule ts_bool_True, assumption)*
 apply *simp+*

apply (*case_tac lose t = [False]*)
apply *simp*
apply (*case_tac a1 t = [sc_ack]*)
apply *simp*
apply (*subgoal_tac a1 t = []*)
  prefer *2*
  apply (*erule aType_empty, assumption*)
apply *simp+*
apply (*subgoal_tac lose t = [True]*)
  prefer *2*
  apply (*erule ts_bool_True, assumption*)
apply *simp+*
apply (*case_tac stop t = [stop_vc]*)
apply *simp*
apply (*subgoal_tac stop t = []*)
  prefer *2*
  apply (*erule stopType_empty, assumption*)
apply *simp*
apply (*case_tac lose t = [False]*)
apply *simp*
apply (*subgoal_tac lose t = [True]*)
  prefer *2*
  apply (*erule ts_bool_True, assumption*)
apply *simp*
done

lemma *tiTable_ack_st_hd*:
[[*tiTable_SampleT req a1 stop lose st_in b ack i1 vc st_out*;
  ts lose; *msg (Suc 0) a1*; *msg (Suc 0) stop*]]
  ⟹ *st_out t = hd (ack t)*
by (*simp add: tiTable_ack_st*)

lemma *tiTable_i1_1*:
[[*tiTable_SampleT req a1 stop lose st_in b ack i1 vc st_out*;
  ts lose; *msg (Suc 0) a1*; *msg (Suc 0) stop*;
  ack t = [connection_ok]]]
  ⟹ *i1 t = []*
apply (*simp add: tiTable_SampleT_def*)
apply (*erule_tac x=t in allE*)
apply *clarify*
apply (*case_tac st_in t*)
apply *simp+*
apply (*subgoal_tac lose t = [False]*)
  prefer *2*
  apply (*erule ts_bool_False, assumption*)
apply *simp+*
apply (*case_tac req t = [send]*)
apply *simp+*
apply (*subgoal_tac lose t = [True]*)
  prefer *2*
  apply (*erule ts_bool_True, assumption*)
apply *simp+*
apply (*subgoal_tac lose t = [False]*)
  prefer *2*
  apply (*erule ts_bool_False, assumption*)

apply *simp+*
apply (*case_tac a1 t = [sc_ack]*)
apply *simp*
apply (*subgoaL_tac lose t = [True]*)
  prefer *2*
  apply (*erule ts_booL True, assumption*)
apply *simp+*
apply (*subgoaL_tac a1 t = []*)
  prefer *2*
  apply (*erule aType_empty, assumption*)
apply *simp+*
apply (*subgoaL_tac lose t = [True]*)
  prefer *2*
  apply (*erule ts_booL True, assumption*)
apply *simp+*
apply (*case_tac stop t = [stop_vc]*)
apply *simp*
apply (*subgoaL_tac stop t = []*)
  prefer *2*
  apply (*erule stopType_empty, assumption*)
apply *simp*
apply (*subgoaL_tac lose t = [True]*)
  prefer *2*
  apply (*erule ts_booL True, assumption*)
apply *simp*
done

lemma *tiTable_i1_2:*
[*tiTable_Sample T req a1 stop lose st_in b ack i1 vc st_out;*
  ts lose; msg (Suc 0) a1; msg (Suc 0) stop;
  ack t = [call]]
  $\implies$ *i1 t = []*
apply (*simp add: tiTable_Sample T_def*)
apply (*erule_tac x=t in allE*)
apply *clarify*
apply (*case_tac st_in t*)
apply *simp+*
apply (*subgoaL_tac lose t = [True]*)
  prefer *2*
  apply (*erule ts_booL True, assumption*)
apply *simp+*
apply (*case_tac req t = [send]*)
apply *simp+*
apply (*subgoaL_tac lose t = [True]*)
  prefer *2*
  apply (*erule ts_booL True, assumption*)
apply *simp+*
apply (*subgoaL_tac lose t = [False]*)
  prefer *2*
  apply (*erule ts_booL False, assumption*)
apply *simp+*
apply (*case_tac a1 t = [sc_ack]*)
apply *simp*
apply (*subgoaL_tac lose t = [True]*)
  prefer *2*
  apply (*erule ts_booL True, assumption*)

apply *simp+*
apply (*subgoal_tac a1 t = []*)
  prefer *2*
  apply (*erule aType_empty, assumption*)
apply *simp+*
apply (*subgoal_tac lose t = [True]*)
  prefer *2*
  apply (*erule ts_bool_True, assumption*)
apply *simp+*
apply (*case_tac stop t = [stop_vc]*)
apply *simp*
apply (*subgoal_tac stop t = []*)
  prefer *2*
  apply (*erule stopType_empty, assumption*)
apply *simp*
apply (*subgoal_tac lose t = [True]*)
  prefer *2*
  apply (*erule ts_bool_True, assumption*)
apply *simp*
done

lemma *tiTable_ack_init*:
⟦ *tiTable_SampleT req a1 stop lose*
              (*fin_inf_append [init_state] st*)
                 b ack i1 vc st;
 ts lose; *msg (Suc 0) a1*; *msg (Suc 0) stop*;
 ∀ *t1* ≤ *t. req t1 = []* ⟧
⟹ *ack t = [init_state]*
apply (*induct t*)
apply (*simp add: tiTable_SampleT_def*)
apply (*erule_tac x=0 in allE*)
apply *clarify*
apply (*simp add: fin_inf_append_def*)

apply *simp*
apply (*subgoal_tac st t = hd (ack t)*)
  prefer *2*
  apply (*simp add: tiTable_ack_st_hd*)
apply (*subgoal_tac*
 (*fin_inf_append [init_state] st*) (*Suc t*) = *init_state*)
  prefer *2*
  apply (*simp add: correct_fin_inf_append2*)
apply (*simp add: tiTable_SampleT_def*)
done

lemma *tiTable_i1_3*:
⟦ *tiTable_SampleT req a1 stop lose*
              (*fin_inf_append [init_state] st*)
                 b ack i1 vc st;
 ts lose; *msg (Suc 0) a1*; *msg (Suc 0) stop*;
 ∀ *t1* ≤ *t. req t1 = []* ⟧
⟹ *i1 t = []*
apply (*subgoal_tac ack t = [init_state]*)
  prefer *2*
  apply (*simp add: tiTable_ack_init*)

apply (*subgoal_tac st t = hd (ack t)*)
  prefer *2*
  apply (*simp add: tiTable_ack_st_hd*)
apply (*subgoal_tac*
  (*fin_inf_append* [*init_state*] *st*) (*Suc t*) = *init_state*)
  prefer *2*
  apply (*simp add: correct_fin_inf_append2*)
apply (*simp add: tiTable_SampleT_def*)
apply (*erule_tac x=t in allE*)
apply *clarify*
apply (*case_tac fin_inf_append* [*init_state*] *st t*)
apply *simp+*
apply (*subgoal_tac lose t* = [*True*])
  prefer *2*
  apply (*erule ts_bool_True, assumption*)
apply *simp+*
apply (*subgoal_tac lose t* = [*True*])
  prefer *2*
  apply (*erule ts_bool_True, assumption*)
apply *simp+*
apply (*case_tac a1 t* = [*sc_ack*])
apply *simp*
apply (*subgoal_tac lose t* = [*True*])
  prefer *2*
  apply (*erule ts_bool_True, assumption*)
apply *simp+*
apply (*subgoal_tac a1 t* = [])
  prefer *2*
  apply (*erule aType_empty, assumption*)
apply *simp+*
apply (*subgoal_tac lose t* = [*True*])
  prefer *2*
  apply (*erule ts_bool_True, assumption*)
apply *simp+*
apply (*case_tac stop t* = [*stop_vc*])
apply *simp*
apply (*subgoal_tac stop t* = [])
  prefer *2*
  apply (*erule stopType_empty, assumption*)
apply *simp*
apply (*subgoal_tac lose t* = [*True*])
  prefer *2*
  apply (*erule ts_bool_True, assumption*)
apply *simp*
done

lemma *tiTable_i1_4* :
$\llbracket$ *tiTable_SampleT req a1 stop lose*
            (*fin_inf_append* [*init_state*] *st*)
              b ack i1 vc st;
  ts lose; *msg* (*Suc 0*) *a1*; *msg* (*Suc 0*) *stop*;
  $\forall$ *t1* $\leq$ *t. req t1* = []; *req* (*Suc t*) = [*init*];
  $\forall$ *m* < *k* + *3. req* (*t* + *m*) $\neq$ [*send*];
  $\forall$ *m* $\leq$ *k. ack* (*Suc* (*Suc* (*t* + *m*))) = [*connection_ok*];
  $\forall$ *j* $\leq$ *k* + *3. lose* (*t* + *j*) = [*False*] $\rrbracket$
  $\implies$ $\forall$ *t2* < (*t* + *3* + *k*). *i1 t2* = []

```
apply clarify
apply (case_tac t2 ≤ t)
apply (simp add: tiTable_i1_3)

 apply (subgoal_tac ack t = [init_state])
   prefer 2
   apply (simp add: tiTable_ack_init)
 apply (subgoal_tac st t = hd (ack t))
   prefer 2
   apply (simp add: tiTable_ack_st_hd)
 apply (subgoal_tac
  (fin_inf_append [init_state] st) (Suc t) = init_state)
   prefer 2
   apply (simp add: correct_fin_inf_append2)
 apply (subgoal_tac st (Suc t) = call)
  prefer 2
  apply (simp add: tiTable_SampleT_def)
 apply (case_tac t2 = Suc t)
 apply (simp add: tiTable_SampleT_def)

 apply (subgoal_tac
(fin_inf_append [init_state] st) (Suc (Suc t)) = call)
  prefer 2
  apply (simp add: correct_fin_inf_append2)
 apply (subgoal_tac st (Suc (Suc t)) = connection_ok)
  prefer 2
  apply (simp add: tiTable_SampleT_def)
 apply (erule_tac x=Suc (Suc t) in allE)
 apply clarify
 apply simp
 apply (case_tac lose (Suc (Suc t)) = [False])
 apply simp+
 apply (subgoal_tac lose (Suc (Suc t)) = [True])
   prefer 2
   apply (erule ts_bool_True, assumption)
 apply simp
 apply (rotate_tac 7)
 apply (erule_tac x=Suc (Suc 0) in allE)
 apply simp
 apply (case_tac t2 = Suc (Suc t))
 apply (subgoal_tac ack (Suc (Suc t)) = [st (Suc (Suc t))])
   prefer 2
   apply (erule tiTable_ack_st, assumption+)
 apply (simp add: tiTable_i1_1)
 apply (subgoal_tac Suc (Suc t) < t2)
   prefer 2
   apply simp
 apply (subgoal_tac ack t2 = [connection_ok])
   prefer 2
   apply (erule aux_ack_t2, assumption+)
 apply (simp add: tiTable_i1_1)
 done


lemma tiTable_ack_ok:
 ⟦ ∀j≤ d + 2. lose (t + j) = [False];
  ts lose; msg (Suc 0) a; msg (Suc 0) stop;
```

req (*Suc t*) ≠ [*send*];
ack t = [*connection_ok*];   *msg* (*Suc 0*) *a1*;
tiTable_SampleT req a1 stop lose (*fin_inf_append* [*init_state*] *st*) *b ack i1 vc st*⟧
⟹ *ack* (*Suc t*) = [*connection_ok*]
apply (*subgoal_tac st t = hd* (*ack t*))
  prefer *2*
  apply (*simp add*: *tiTable_ack_st_hd*)
apply (*subgoal_tac*
  (*fin_inf_append* [*init_state*] *st*) (*Suc t*) = *connection_ok*)
  prefer *2*
  apply (*simp add*: *correct_fin_inf_append2*)
apply (*simp add*: *tiTable_SampleT_def*)
apply (*rotate_tac* −3)
apply (*erule_tac x=Suc t* **in** *allE*)
apply *clarify*
apply (*erule_tac x=Suc 0* **in** *allE*)
apply *simp*
done

lemma *Gateway_L7a*:
⟦ ∀*j*≤ *d* + 2. *lose* (*t* + *j*) = [*False*];
  Gateway req dt a stop lose d ack i vc; *ts lose*;
  msg (*Suc 0*) *a*; *msg* (*Suc 0*) *stop*;
  msg (*Suc 0*) *req*; *req* (*Suc t*) ≠ [*send*];
  ack (*t*) = [*connection_ok*]⟧
⟹ *ack* (*Suc t*) = [*connection_ok*]
apply (*simp add*: *Gateway_def*)
apply *clarify*
apply (*simp add*: *Sample_def*)
apply (*subgoal_tac msg* (*Suc 0*) *a1*)
  prefer *2*
  apply (*simp add*: *Loss_Delay_msg_a*)
apply *clarify*
apply (*simp add*: *Sample_L_def*)
apply *clarify*
apply (*erule_tac*
  V=∀ *t. buffer t* =
  (*if dt t* = [] *then fin_inf_append* [[]] *buffer t else dt t*)
  in *thin_rl*)
apply (*simp add*: *tiTable_ack_ok*)
done

lemma *Sample_L_buffer*:
⟦ *Sample_L req dt a1 stop lose*
        (*fin_inf_append* [*init_state*] *st*)
        (*fin_inf_append* [[]] *buffer*)
        ack i1 vc st buffer ⟧
⟹ *buffer t* = *inf_last_ti dt t*
apply (*simp only*: *Sample_L_def*)
apply *clarify*
apply (*erule_tac*
  V=*tiTable_SampleT req a1 stop lose*
  (*fin_inf_append* [*init_state*] *st*)
  (*fin_inf_append* [[]] *buffer*) *ack i1 vc st*
  in *thin_rl*)

apply (*subgoal_tac*
  ∀ *t. buffer t =*
  (*if dt t = [] then fin_inf_append [[]] buffer t else dt t*))
  prefer *2*
  apply *simp*
apply (*erule_tac x=t in allE*)
apply (*induct t*)
apply (*simp add: fin_inf_append_def*)
apply *atomize*
apply *simp*
apply (*case_tac dt t = []*)
apply (*simp add: correct_fin_inf_append1*)
apply *simp*
apply (*erule_tac x=t in allE*)
apply (*simp add: correct_fin_inf_append1*)
done

lemma *Sample_L_i1_buffer*:
⟦ *msg (Suc 0) req*; *msg (Suc 0) a*; *msg (Suc 0) stop*;
  ack t = [connection_ok]; *req (Suc t) = [send]*;
  ∀ *k≤Suc d. lose (t + k) = [False]*;
  ts lose; *msg (Suc 0) a1*;
  Sample_L req dt a1 stop lose
      (*fin_inf_append [init_state] st*)
      (*fin_inf_append [[]] buffer*) *ack i1 vc st buffer*⟧
  ⟹ *i1 (Suc t) = buffer t*
apply (*subgoal_tac buffer t = inf_last_ti dt t*)
  prefer *2*
  apply (*simp add: Sample_L_buffer*)
apply (*simp add: Sample_L_def*)
apply *clarify*
apply (*subgoal_tac st t = hd (ack t)*)
  prefer *2*
  apply (*simp add: tiTable_ack_st_hd*)
apply (*subgoal_tac*
  (*fin_inf_append [init_state] st*) (*Suc t*) *= connection_ok*)
  prefer *2*
  apply (*simp add: correct_fin_inf_append1*)
apply (*simp add: tiTable_SampleT_def*)
apply (*rotate_tac −3*)
apply (*erule_tac x=Suc t in allE*)
apply *clarify*
apply *simp*
apply (*case_tac lose (Suc t) = [True]*)
apply *simp*
apply (*erule_tac x=Suc 0 in allE*)
apply *simp*
apply *simp*
apply (*subgoal_tac lose (Suc t) = [False]*)
  prefer *2*
  apply (*simp add: ts_bool_False*)
apply (*simp add: correct_fin_inf_append1*)
done

lemma *Sample_sending_data*:

⟦*msg (Suc 0) stop*; *ts lose*; *msg (Suc 0) req*;
 msg (Suc 0) a1;
 $\forall j {\le} 2 * d.\ lose\ (t + j) = [False]$;
 ack t = [sending_data];
 Sample req dt a1 stop lose ack i1 vc;
 $x \le d + d$;
 $\forall t4 \le t + d + d.\ a1\ t4 = []$⟧
 $\implies ack\ (t + x) = [sending\_data]$
 apply (*simp add: Sample_def*)
 apply *clarify*
 apply (*simp add: Sample_L_def*)
 apply *clarify*
 apply (*erule_tac*
 $V {=} \forall t.\ buffer\ t =$
 (*if dt t = [] then fin_inf_append [[]] buffer t else dt t*)
 in thin_rl)
 apply (*induct x*)
 apply *simp*

 apply *atomize*
 apply (*erule_tac x=st in allE*)
 apply (*rotate_tac −1*)
 apply (*erule_tac x=buffer in allE*)
 apply *clarify*
 apply *simp*
 apply (*subgoal_tac st (t + x) = hd (ack (t + x))*)
  prefer *2*
  apply (*simp add: tiTable_ack_st_hd*)
 apply (*subgoal_tac*
 (*fin_inf_append [init_state] st*) (*Suc (t + x)*) = *sending_data*)
  prefer *2*
  apply (*simp add: fin_inf_append_def*)
 apply (*erule_tac x=Suc x in allE*)
 apply (*simp only: tiTable_SampleT_def*)
 apply (*erule_tac x=Suc (t + x) in allE*)
 apply (*subgoal_tac*
 $Suc\ (t + x) \le 2 * d + t$)
  prefer *2*
  apply *simp*
 apply *simp*
 done

lemma *ServiceCenter_a_msg*:
ServiceCenter i a $\implies$ *msg (Suc 0) a*
 apply (*simp add: ServiceCenter_def msg_def*)
 apply *clarify*
 apply (*case_tac t*)
 apply (*simp split add: split_if_asm*)+
 done
end

B.7. Proof of the Automotive-Gateway System Properties

theory *Gateway_proof = Main + Gateway + ts_bool_stream*

+ Gateway_proof_aux:

B.7.1. Properties of the Gateway

lemma *Gateway_L1*:

⟦ *Gateway req dt a stop lose d ack i vc;*
  msg (Suc 0) req; msg (Suc 0) a; msg (Suc 0) stop;
  ack t = [init_state]; req (Suc t) = [init];
  ts lose;
  lose (Suc t) = [False]; lose (Suc (Suc t)) = [False]⟧
⟹ *ack (Suc (Suc t)) = [connection_ok]*
  apply (*simp add: Gateway_def*)
  apply *clarify*
  apply (*simp add: Sample_def*)
  apply (*subgoal_tac msg (Suc 0) a1*)
    prefer *?*
    apply (*simp add: Loss_Delay_msg_a*)
  apply (*simp add: Sample_L_def*)
  apply *clarify*
  apply (*erule_tac*
    V=∀ t. buffer t =
    (if dt t = [] then fin_inf_append [[]] buffer t else dt t)
    in *thin_rl*)
  apply (*subgoal_tac st t = hd (ack t)*)
    prefer *2*
    apply (*simp add: tiTable_ack_st_hd*)
  apply (*subgoal_tac (fin_inf_append [init_state] st) (Suc t) = init_state*)
    prefer *2*
    apply (*simp add: correct_fin_inf_append1*)
  apply (*subgoal_tac st (Suc t) = call*)
    prefer *2*
    apply (*simp add: tiTable_SampleT_def*)
  apply (*subgoal_tac (fin_inf_append [init_state] st) (Suc (Suc t)) = call*)
    prefer *2*
    apply (*simp add: correct_fin_inf_append1*)
  apply (*simp add: tiTable_SampleT_def*)
  done

lemma *Gateway_L2*:

⟦ *Gateway req dt a stop lose d ack i vc;*
  msg (Suc 0) req; msg (Suc 0) a; msg (Suc 0) stop;
  ack t = [connection_ok]; req (Suc t) = [send];
  ∀ k≤Suc d. lose (t + k) = [False]; ts lose ⟧
⟹ *i (Suc (t + d)) = inf_last_ti dt t*
  apply (*simp add: Gateway_def*)
  apply *clarify*
  apply (*simp add: Sample_def*)
  apply (*subgoal_tac msg (Suc 0) a1*)
    prefer *2*
    apply (*simp add: Loss_Delay_msg_a*)
  apply *clarify*
  apply (*subgoal_tac buffer t = inf_last_ti dt t*)
    prefer *2*
    apply (*simp add: Sample_L_buffer*)
  apply (*subgoal_tac i1 (Suc t) = buffer t*)
    prefer *2*

    apply (*simp add: Sample_L_i1_buffer*)
apply (*subgoal_tac i2 ((Suc t) + d) = i1 (Suc t)*)
   prefer *2*
   apply (*simp add: Delay_def*)
   apply (*rotate_tac 6*)
   apply (*erule_tac x=Suc t in allE*)
   apply *simp*
apply (*subgoal_tac i ((Suc t) + d) = i2 ((Suc t) + d)*)
   prefer *2*
   apply (*simp add: Loss_def*)
   apply (*erule_tac x=Suc d in allE*)
   apply (*erule_tac x=(Suc t) + d in allE*)
   apply *simp*
apply *simp*
done

lemma *Gateway_L3*:
⟦ *Gateway req dt a stop lose d ack i vc*;
  msg (Suc 0) req; *msg (Suc 0) a*; *msg (Suc 0) stop*;
  ts lose; *ack t = [connection_ok]*;
  req (Suc t) = [send]; *∀ k≤Suc d. lose (t + k) = [False]*⟧
⟹ *ack (Suc t) = [sending_data]*
apply (*simp add: Gateway_def*)
apply *clarify*
apply (*simp add: Sample_def*)
apply (*subgoal_tac msg (Suc 0) a1*)
  prefer *2*
  apply (*simp add: Loss_Delay_msg_a*)
apply (*simp add: Sample_L_def*)
apply *clarify*
apply (*subgoal_tac st t = hd (ack t)*)
  prefer *2*
  apply (*simp add: tiTable_ack_st_hd*)
apply (*erule_tac
  V =∀ t. buffer t =
  (if dt t = [] then fin_inf_append [[]] buffer t else dt t)
  in thin_rl*)
apply (*subgoal_tac (fin_inf_append [init_state] st) (Suc t) = connection_ok*)
  prefer *2*
  apply (*simp add: correct_fin_inf_append1*)
apply (*subgoal_tac st (Suc t) = sending_data*)
  prefer *2*
  apply (*simp add: tiTable_SampleT_def*)
  apply (*erule_tac x=Suc 0 in allE*)
  apply (*erule_tac x=Suc t in allE*)
  apply *simp*
apply (*subgoal_tac ack (Suc t) = [st (Suc t)]*)
  prefer *2*
  apply (*erule tiTable_ack_st, assumption+*)
apply *simp*
done

lemma *Gateway_L4*:
⟦ *Gateway req dt a stop lose d ack i vc*;
  msg (Suc 0) req; *msg (Suc 0) a*; *msg (Suc 0) stop*;

>   *ts lose; ack (t + d) = [sending_data];*
>   *a (Suc t) = [sc_ack];*
> $\forall\, k \leq Suc\ d.\ lose\ (t + k) = [False]$]]
> $\Longrightarrow vc\ (Suc\ (t + d)) = [vc\_com]$
> **apply** (*simp add: Gateway_def*)
> **apply** *clarify*
> **apply** (*simp add: Sample_def*)
> **apply** (*subgoal_tac msg (Suc 0) a1*)
>   **prefer** *2*
>   **apply** (*simp add: Loss_Delay_msg_a*)
> **apply** *clarify*
> **apply** (*simp add: Sample_L_def*)
> **apply** *clarify*
> **apply** (*subgoal_tac st (t+d) = hd (ack (t+d))*)
>   **prefer** *2*
>   **apply** (*simp add: tiTable_ack_st_hd*)
> **apply** (*subgoal_tac (fin_inf_append [init_state] st) (Suc (t + d)) = sending_data*)
>   **prefer** *2*
>   **apply** (*simp add: correct_fin_inf_append1*)
> **apply** (*subgoal_tac a2 (Suc t) = a (Suc t)*)
>   **prefer** *2*
>   **apply** (*simp add: Loss_def*)
>   **apply** (*erule_tac x=Suc 0 in allE*)
>   **apply** (*erule_tac x=(Suc t) in allE*)
>   **apply** *simp*
> **apply** (*subgoal_tac a1 ((Suc t) + d) = a2 (Suc t)*)
>   **prefer** *2*
>   **apply** (*simp add: Delay_def*)
>   **apply** (*rotate_tac 7*)
>   **apply** (*erule_tac x=Suc t in allE*)
>   **apply** *simp*
> **apply** (*simp add: tiTable_SampleT_def*)
> **apply** (*rotate_tac −5*)
> **apply** (*erule_tac x=Suc (t+d) in allE*)
> **apply** *simp*
> **apply** (*erule_tac x=Suc d in allE*)
> **apply** *simp*
> **done**

lemma *Gateway_L5:*
[[*Gateway req dt a stop lose d ack i vc;*
   msg (Suc 0) req; msg (Suc 0) a; msg (Suc 0) stop;
   $\forall\ j \leq Suc\ d.\ a\ (t+j) = [];$
   ack (t+d) = [sending_data];
   $\forall\, k \leq (d+d).\ lose\ (t + k) = [False];\ ts\ lose$]]
$\Longrightarrow j \leq d \longrightarrow ack\ (t+d+j) = [sending\_data]$
apply (*simp add: Gateway_def*)
apply (*induct j*)
apply *simp*
apply *atomize*
apply *simp*
apply *auto*
apply (*simp add: Sample_def*)
apply (*subgoal_tac msg (Suc 0) a1*)
  prefer *2*
  apply (*simp add: Loss_Delay_msg_a*)

apply *clarify*
apply (*erule_tac x=Suc j* **in** *allE*)

apply (*simp add: Loss_def*)
apply (*subgoal_tac* ∀ *k*≤*d* + *d. lose* (*t* + *k*) = [*False*])
  prefer *2*
  apply *simp*
apply (*subgoal_tac d* + *Suc j* ≤ *d* + *d*)
  prefer *2*
  apply *simp*
apply (*simp add: Delay_def*)
apply (*simp add: Sample_L_def*)
apply *clarify*
apply (*erule_tac*
  V=∀ *t. buffer t* =
  (*if dt t* = [] *then fin_inf_append* [[]] *buffer t else dt t*)
  in *thin_rl*)
apply (*subgoal_tac st* (*t+d+j*) = *hd* (*ack* (*t+d+j*)))
  prefer *2*
  apply (*simp add: tiTable_ack_st_hd*)
apply (*subgoal_tac*
  (*fin_inf_append* [*init_state*] *st*) (*Suc* (*t+d+j*)) = *sending_data*)
  prefer *2*
  apply (*simp add: correct_fin_inf_append1*)
apply (*subgoal_tac Suc* (*d* + *j*) ≤ *d* + *d*)
  prefer *2*
  apply *simp*
apply (*subgoal_tac st t* = *hd* (*ack t*))
  prefer *2*
  apply (*simp add: tiTable_ack_st_hd*)
apply (*simp add: tiTable_SampleT_def*)
apply (*erule_tac x=Suc* (*t* + *j*) **in** *allE*)
apply (*erule_tac x=Suc* (*t* + *j*) **in** *allE*)
apply (*erule_tac x=Suc* (*d* + *j*) **in** *allE*)
apply (*erule_tac x=Suc* (*t* + *d* + *j*) **in** *allE*)
apply *simp*
apply (*subgoal_tac* (*Suc* (*t* + *j* + *d*)) = *Suc* (*t* + *d* + *j*))
  prefer *2*
  apply *simp*
apply *simp*
apply (*subgoal_tac* (*Suc* (*t* + (*d* + *j*))) = *Suc* (*t* + *d* + *j*))
  prefer *2*
  apply *simp*
apply (*simp split add: split_if_asm*)
done

lemma *Gateway_L6_induction*:
⟦ *msg* (*Suc 0*) *req; ts lose; msg* (*Suc 0*) *a;*
  msg (*Suc 0*) *stop;*
  ∀ *j*≤ *k. lose* (*t* + *j*) = [*False*];
  ∀ *m* ≤ *k. req* (*t* + *m*) ≠ [*send*]; *ack t* = [*connection_ok*];
  Sample req dt a1 stop lose ack i1 vc;
  Delay a2 i1 d a1 i2; Loss lose a i2 a2 i; m ≤ *k*⟧
  ⟹ *ack* (*t* + *m*) = [*connection_ok*]
  apply (*induct m*)
  apply (*rotate_tac 5*)

apply (*erule_tac x=0* **in** *allE*)
apply *simp*

apply (*simp add: Sample_def*)
apply (*subgoal_tac msg* (*Suc 0*) *a1*)
  prefer *2*
  apply (*simp add: Loss_Delay_msg_a*)
apply *clarify*
apply (*simp add: Sample_L_def*)
apply *clarify*
apply (*erule_tac*
  $V=\forall t.\ buffer\ t =$
  (*if dt t* = [] *then fin_inf_append* [[]] *buffer t else dt t*)
  in *thin_rl*)
apply (*subgoal_tac st* (*t + m*) = *hd* (*ack* (*t + m*)))
  prefer *2*
  apply (*simp add: tiTable_ack_st_hd*)
apply (*subgoal_tac*
  (*fin_inf_append* [*init_state*] *st*) (*Suc* (*t + m*)) = *connection_ok*)
  prefer *2*
  apply (*simp add: fin_inf_append_def*)
apply (*simp only: tiTable_SampleT_def*)
apply (*rotate_tac* −*3*)
apply (*erule_tac x=Suc* (*t + m*) **in** *allE*)
apply *simp*
apply *clarify*
apply (*erule_tac x=Suc m* **in** *allE*)
apply (*rotate_tac 6*)
apply (*erule_tac x=Suc m* **in** *allE*)
apply *simp*
done

lemma *Gateway_L6*:
⟦ *Gateway req dt a stop lose d ack i vc*;
  msg (*Suc 0*) *req*; *ts lose*; *msg* (*Suc 0*) *a*;
  msg (*Suc 0*) *stop*;
  $\forall j{\leq}k.\ lose\ (t + j) = [False]$;
  $\forall m \leq k.\ req\ (t + m) \neq [send]$; $ack\ t = [connection\_ok]$⟧
      $\Longrightarrow\ \forall m \leq k.\ ack\ (t + m) = [connection\_ok]$
apply (*simp add: Gateway_def*)
apply *clarify*
apply (*simp add: Gateway_L6_induction*)
done

lemma *Gateway_L7*:
⟦ *Gateway req dt a stop lose d ack i vc*;
  ts lose; *msg* (*Suc 0*) *a*; *msg* (*Suc 0*) *stop*;
  msg (*Suc 0*) *req*;
  $\forall\ t1 \leq t.\ req\ t1 = []$; $req\ (Suc\ 0) = [init]$;
  $\forall m < (k + 3).\ req\ (t + m) \neq [send]$;
  $req\ (t + 3 + k) = [send]$; $ack\ t = [init\_state]$;
  $\forall j{\leq}k + d + 3.\ lose\ (t + j) = [False]$ ⟧
      $\Longrightarrow \forall\ t2 < (t + 3 + k + d).\ i\ t2 = []$
  apply (*subgoal_tac*
  $ack\ (Suc\ (Suc\ t)) = [connection\_ok]$)

prefer *2*
apply (*subgoal_tac*
  $\forall j \leq k + d + 3.\ lose\ (t + j) = [False]$)
  prefer *2*
  apply *simp*
apply (*rotate_tac* *−1*)
apply (*erule_tac* *x=Suc 0* **in** *allE*)
apply (*rotate_tac* *−2*)
apply (*erule_tac* *x=Suc (Suc 0)* **in** *allE*)
apply *simp*
apply (*simp add: Gateway_L1*)
apply (*subgoal_tac*
  $\forall m \leq k.\ ack\ ((t + 2) + m) = [connection\_ok]$)
  prefer *2*
  apply (*erule Gateway_L6*)
  apply *assumption+*
  apply *clarify*
  apply (*rotate_tac* *−3*)
  apply (*erule_tac* *x=2+j* **in** *allE*)
  apply *simp*
  apply *clarify*
  apply (*rotate_tac 6*)
  apply (*erule_tac* *x=2+m* **in** *allE*)
  apply *simp*
  apply *simp*
  apply (*subgoal_tac ack (t + 2 + k) = [connection_ok]*)
  prefer *2*
  apply (*rotate_tac* *−1*)
  apply (*erule_tac* *x=k* **in** *allE*)
  apply *simp*
apply (*simp add: Gateway_def*)
apply *clarify*
apply (*simp add: Sample_def*)
apply (*subgoal_tac msg (Suc 0) a1*)
  prefer *2*
  apply (*simp add: Loss_Delay_msg_a*)
apply *clarify*
apply (*simp add: Sample_L_def*)
apply *clarify*
apply (*erule_tac*
  $V = \forall t.\ buffer\ t =$
  (*if dt t = [] then fin_inf_append [[]] buffer t else dt t*)
  in *thin_rl*)
apply (*subgoal_tac* $\forall\ t1 < (t + 3 + k).\ i1\ t1 = []$)
  prefer *2*
  apply (*simp add: tiTable_i1_4*)
apply (*simp add: Delay_def Loss_def*)
apply (*rotate_tac* *−5*)
apply (*erule_tac* *x=t2* **in** *allE*)
apply (*case_tac lose t2 = [False]*)
apply (*simp split add: split_if_asm*)
apply (*rotate_tac* *−3*)
apply (*case_tac t2 < d*)
apply *simp*
apply (*erule_tac* *x=t2 − d* **in** *allE*)
apply *simp*
apply (*subgoal_tac t2 − d < t + 3 + k*)

   prefer *2*
    apply *arith*
   apply (*case_tac t2 − d < d*)
   apply *simp*
   apply (*erule_tac x=t2 − d in allE*)
   apply *simp*
   apply (*simp split add: split_if_asm*)
   done

lemma *Gateway_L8*:
⟦ *Gateway req dt a stop lose d ack i vc*;
  msg (Suc 0) req; *msg (Suc 0) stop*; *ts lose*;
  msg (Suc 0) a;
  ∀ *j≤2 ∗ d. lose (t + j) = [False]*;
  ack t = [sending_data];
  ∀ *t3 ≤ t + d. a t3 = []* ⟧
  ⟹ ∀ *x ≤ d + d. ack (t + x) = [sending_data]*
  apply (*simp add: Gateway_def*)
  apply *clarify*
  apply (*subgoal_tac ∀ t3 ≤ t + d. a2 t3 = []*)
   prefer *2*
   apply *clarify*
   apply (*simp add: Loss_def*)
   apply (*rotate_tac −3*)
   apply (*erule_tac x=t3 in allE*)
   apply (*simp split add: split_if_asm*)
  apply (*subgoal_tac ∀ t4 ≤ t + d + d. a1 t4 = []*)
   prefer *2*
   apply *clarify*
   apply (*simp add: Delay_def*)
   apply (*rotate_tac −5*)
   apply (*case_tac t4 < d*)
   apply (*erule_tac x=t4 in allE*)
   apply *simp*
   apply (*erule_tac x=t4−d in allE*)
   apply (*case_tac t4 − d < d*)
   apply *simp+*
   apply *clarify*
   apply (*rotate_tac 2*)
   apply (*erule_tac x=t4−d in allE*)
   apply (*subgoal_tac t4 − d ≤ t + d*)
    prefer *2*
    apply *arith*
   apply *simp*
  apply (*subgoal_tac msg (Suc 0) a1*)
   prefer *2*
   apply (*simp add: Loss_Delay_msg_a*)
  apply (*simp add: Sample_sending_data*)
  done

B.7.2. Proof of the Refinement Relation for the Gateway Requirements

lemma *Gateway_L0*:
Gateway req dt a stop lose d ack i vc
⟹
GatewayReq req dt a stop lose d ack i vc

apply (*simp add: GatewayReq_def*)
apply *auto*
apply (*simp add: Gateway_L1*)
apply (*simp add: Gateway_L2*)
apply (*simp add: Gateway_L3*)
apply (*simp add: Gateway_L4*)
done

B.7.3. Lemmas about Gateway Requirements

lemma *GatewayReq_L1*:
⟦ *msg (Suc 0) req*; *msg (Suc 0) stop*; *ts lose*; *msg (Suc 0) a*;
  $\forall m \leq k + 2$. *req* $(t + m) \neq [send]$;
  req $(t + 3 + k) = [send]$;
  $\forall j \leq 2 * d + (4 + k)$. *lose* $(t + j) = [False]$;
  GatewayReq req dt a stop lose d ack i vc;
  ack (Suc (Suc t)) = [connection_ok];
  $\forall m \leq k$. *ack* $(t + 2 + m) = [connection\_ok]$⟧
⟹ *ack* $(t + 3 + k) = [sending\_data]$
apply (*simp add: GatewayReq_def*)
apply (*rotate_tac* −3)
apply (*erule_tac x=t+2+k in allE*)
apply *clarify*
apply (*erule_tac x=k in allE*)
apply (*subgoal_tac*
  $\forall ka \leq Suc\ d$. *lose* $(Suc\ (Suc\ (t + k + ka))) = [False]$)
  prefer *2*
  apply (*simp add: aux_lemma_lose_1*)
apply (*simp add: nat_number*)
done

lemma *GatewayReq_L2*:
⟦ *GatewayReq req dt a stop lose d ack i vc*;
  msg (Suc 0) req; *msg (Suc 0) stop*; *ts lose*;
  msg (Suc 0) a;
  req $(t + (3::nat) + k) = [send]$; *inf_last_ti dt t* $\neq []$;
  $\forall j \leq 2 * d + (4 + k)$. *lose* $(t + j) = [False]$;
  ack (Suc (Suc t)) = [connection_ok];
  $\forall m \leq k$. *ack* $(t + 2 + m) = [connection\_ok]$ ⟧
⟹ *i* $(t + 3 + k + d) \neq []$
apply (*rotate_tac* −1)
apply (*erule_tac x=k in allE*)
apply *simp*
apply *subgoal_tac*
  $(\forall (x::nat). x \leq (d+1) \longrightarrow lose\ (t+x) = [False])$)
  prefer *2*
  apply (*simp add: aux_lemma_lose_2*)
apply (*simp add: GatewayReq_def*)
apply (*erule_tac x=(Suc (Suc (t + k))) in allE*)
apply *clarify*
apply (*subgoal_tac*
  $\forall ka \leq Suc\ d$. *lose* $(Suc\ (Suc\ (t + k + ka))) = [False]$)
  prefer *2*
  apply (*simp add: aux_lemma_lose_1*)
apply (*simp add: nat_number*)
apply (*simp split add: split_if_asm*)

apply (*simp add: inf_last_ti_nonempty_k*)
done

B.7.4. Properties of the Gateway System

lemma *GatewaySystem_L1*:
⟦*Gateway req dt a stop lose d ack i vc*;
 ServiceCenter i a;
 GatewayReq req dt a stop lose d ack i vc;
 msg (Suc 0) req; *msg (Suc 0) stop*; *ts lose*;
 msg (Suc 0) a;
 req (t + 3 + k) = [send];
 ∀*j*≤2 * *d* + (*4* + *k*). *lose (t + j) = [False]*;
 ack (t + 3 + k) = [sending_data]; *i (t + 3 + k + d) ≠ []*;
 ∀*t2*<*t* + *3* + *k* + *d*. *i t2 = []*;
 ∀*t3*≤*t* + *3* + *k* + *d*. *a t3 = []*;
 ∀*x* ≤ *d* + *d*. *ack (t + 3 + k + x) = [sending_data]* ⟧
 ⟹ *vc (2 * d + (t + (4 + k))) = [vc_com]*
apply (*rotate_tac −1*)
apply (*erule_tac x=2 * d in allE*)
apply (*simp add: GatewayReq_def*)
apply (*erule_tac x=(d + (t + (3 + k))) in allE*)
apply *clarify*
apply *simp*
apply (*subgoal_tac a (4 + (d + (t + k))) = [sc_ack]*)
  prefer *2*
  apply (*simp add:ServiceCenter_def*)
  apply (*erule_tac x=(t + 3 + k + d) in allE*)
  apply (*subgoal_tac*
    4 + (t + (k + d)) = 4 + (d + (t + k)))
    prefer *2*
    apply *simp*
  apply *simp*
apply *simp*
 apply (*subgoal_tac*
   *(t + 3 + k + 2 * d) = (2 * d + (t + (3 + k)))*)
   prefer *2*
   apply *simp*
 apply *simp*
apply (*subgoal_tac*
  ∀*ka*≤*Suc d*. *lose (d + (t + (3 + k)) + ka) = [False]*)
  prefer *2*
  apply (*simp add: aux_lemma_lose_3*)
apply *simp*
apply (*subgoal_tac*
  *(4 + (2 * d + (t + k))) = (2 * d + (t + (4 + k)))*)
  prefer *2*
  apply *arith*
apply *simp*
done

lemma *GatewaySystem_L2*:
⟦*Gateway req dt a stop lose d ack i vc*;
 ServiceCenter i a;
 GatewayReq req dt a stop lose d ack i vc;
 msg (Suc 0) req; *msg (Suc 0) stop*;

ts lose; *msg (Suc 0) a*;
ack t = [*init_state*]; *req (Suc t)* = [*init*];
∀ *t1* ≤ *t. req t1* = [];
∀ *m* ≤ *k* + *2. req (t* + *m)* ≠ [*send*]; *req (t* + *3* + *k)* = [*send*];
inf_last_ti dt t ≠ [];
∀ *j* ≤ *2* * *d* + (*4* + *k*). *lose (t* + *j)* = [*False*]]
⟹ *vc (2* * *d* + (*t* + (*4* + *k*))) = [*vc_com*]
apply (*subgoal_tac ack (Suc (Suc t))* = [*connection_ok*])
  prefer *2*
  apply (*erule_tac*
    V=∀ *m* ≤ *k* + *2. req (t* + *m)* ≠ [*send*]
    in *thin_rl*)
  apply (*simp add: GatewayReq_def*)
  apply (*rotate_tac 2*)
  apply (*erule_tac x*=*t* **in** *allE*)
  apply *simp*
  apply (*subgoal_tac*
  ∀ *j* ≤ *2* * *d* + (*4* + *k*). *lose (t* + *j)* = [*False*])
    prefer *2*
    apply *simp*
  apply (*rotate_tac* −*1*)
  apply (*erule_tac x*=*Suc 0* **in** *allE*)
  apply (*rotate_tac 9*)
  apply (*erule_tac x*=*Suc (Suc 0)* **in** *allE*)
  apply *simp*

apply (*subgoal_tac* ∀ *m* ≤ *k. ack (t* + *2* + *m)* = [*connection_ok*])
  prefer *2*
  apply (*subgoal_tac* ∀ *j* ≤ *k. lose (t* + *2* + *j)* = [*False*])
    prefer *2*
    apply *clarify*
    apply (*rotate_tac* −*3*)
    apply (*erule_tac x*=*2* + *j* **in** *allE*)
    apply *simp*
  apply (*erule Gateway_L6, assumption*+)
  apply *clarify*
  apply (*rotate_tac 9*)
  apply (*erule_tac x*=*2* + *m* **in** *allE*)
  apply *simp*
  apply *simp*

apply (*subgoal_tac ack (t* + *3* + *k)* = [*sending_data*])
  prefer *2*
  apply (*simp add: GatewayReq_L1*)
apply (*subgoal_tac i (t* + *3* + *k* + *d)* ≠ [])
  prefer *2*
  apply (*simp add: GatewayReq_L2*)

apply (*subgoal_tac* ∀ *t2* < (*t* + *3* + *k* + *d*). *i t2* = [])
  prefer *2*
  apply (*erule Gateway_L7, assumption*+)
  apply *clarify*
  apply (*rotate_tac 9*)
  apply (*erule_tac x*=*m* **in** *allE*)
  apply *simp*
  apply *assumption*+
  apply *clarify*

> **apply** (*rotate_tac −5*)
> **apply** (*erule_tac x=j in allE*)
> **apply** *simp*
> **apply** (*subgoal_tac ∀ t3 ≤ (t + 3 + k + d). a t3 = []*)
>   **prefer** *2*
>   **apply** (*simp add: ServiceCenter_def*)
>   **apply** *clarify*
>   **apply** (*case_tac t3*)
>   **apply** *simp*+
> **apply** (*erule_tac x=Suc (t + 3 + k + d) in allE*)
> **apply** *simp*
> **apply** (*subgoal_tac*
>   ∀x ≤ d + d. ack (t + 3 + k + x) = [sending_data]*)
>   **prefer** *2*
>   **apply** (*erule Gateway_L8, assumption*+)
>   **apply** *clarify*
>   **apply** (*rotate_tac 11*)
>   **apply** (*erule_tac x=3 + k + j in allE*)
>   **apply** (*subgoal_tac t + (3 + k + j) = t + 3 + k + j*)
>   **prefer** *2*
>   **apply** *arith*
>   **apply** *simp*
> **apply** *assumption*+
> **apply** (*simp add: GatewaySystem_L1*)
> **done**

lemma *GatewaySystem_L3*:
⟦ *Gateway req dt a stop lose d ack i vc*;
  ServiceCenter i a; *msg (Suc 0) req*;
  GatewayReq req dt a stop lose d ack i vc;
  msg (Suc 0) stop; *ts lose*; *msg (Suc 0) a*;
  (*dt (Suc t) ≠ [] ∨ dt (Suc (Suc t)) ≠ []*);
  ack t = [init_state]; *req (Suc t) = [init]*;
  ∀t1≤t. req t1 = []; ∀m ≤ k + 2. req (t + m) ≠ [send];
  req (t + 3 + k) = [send];
  ∀j≤2 * d + (4 + k). lose (t + j) = [False]⟧
⟹ *vc (2 * d + (t + (4 + k))) = [vc_com]*
> **apply** (*subgoal_tac ack (Suc (Suc t)) = [connection_ok]*)
>   **prefer** *2*
>   **apply** (*erule_tac*
>   V=∀ m ≤ k + 2. req (t + m) ≠ [send]*
>   **in** *thin_rl*)
>   **apply** (*simp add: GatewayReq_def*)
>   **apply** (*rotate_tac 2*)
>   **apply** (*erule_tac x=t in allE*)
>   **apply** *simp*
>   **apply** (*subgoal_tac*
>   ∀j≤2 * d + (4 + k). lose (t + j) = [False]*)
>   **prefer** *2*
>   **apply** *simp*
>   **apply** (*rotate_tac −1*)
>   **apply** (*erule_tac x=Suc 0 in allE*)
>   **apply** (*rotate_tac 9*)
>   **apply** (*erule_tac x=Suc (Suc 0) in allE*)
>   **apply** *simp*

apply (*subgoaL_tac* ∀ *m* ≤ *k*. *ack* (*t* + *2* + *m*) = [*connection_ok*])
  prefer *2*
  apply (*erule Gateway_L6*, *assumption*+)
  apply *clarify*
  apply (*rotate_tac* −*3*)
  apply (*erule_tac* *x*=*2* + *j* **in** *allE*)
  apply *simp*
  apply *clarify*
  apply (*rotate_tac* −*6*)
  apply (*erule_tac* *x*=*2* + *m* **in** *allE*)
  apply *simp*
  apply *simp*

apply (*subgoaL_tac ack* (*t* + *3* + *k*) = [*sending_data*])
  prefer *2*
  apply (*simp add: GatewayReq_L1*)
apply (*subgoaL_tac i* (*t* + *3* + *k* + *d*) ≠ [])
  prefer *2*
  apply (*simp add: GatewayReq_def*)
  apply (*erule_tac* *x*=*t* + *2* + *k* **in** *allE*)
  apply *simp*
  apply (*subgoaL_tac*
    ∀ *ka*≤*Suc d*. *lose* (*Suc* (*Suc* (*t* + *k* + *ka*))) = [*False*])
    prefer *2*
    apply (*simp add: aux_lemma_lose_1*)

  apply *simp*
  apply (*subgoaL_tac inf_last_ti dt* (*Suc* (*Suc t*)) ≠ [])
    prefer *2*
    apply *simp*
  apply *clarify*
  apply (*subgoaL_tac inf_last_ti dt* (*Suc* (*Suc* (*t*+*k*))) ≠ [])
    prefer *2*
    apply (*erule inf_last_ti2*)
  apply (*subgoaL_tac*
    (*if dt* (*Suc* (*Suc* (*t* + *k*))) ≠ [] *then dt* (*Suc* (*Suc* (*t* + *k*)))
    else inf_last_ti dt (*Suc* (*t* + *k*)))
    = *inf_last_ti dt* (*Suc* (*Suc* (*t* + *k*))))
    prefer *2*
    apply *simp*
  apply (*simp* (*no_asm_use*))
  apply (*simp add: nat_number*)

apply (*subgoaL_tac* ∀ *t2* < (*t* + *3* + *k* + *d*). *i t2* = [])
  prefer *2*
  apply (*erule Gateway_L7*, *assumption*+)
  apply *clarify*
  apply (*rotate_tac 10*)
  apply (*erule_tac* *x*=*m* **in** *allE*)
  apply *simp*
  apply *assumption*+
  apply *clarify*
  apply (*rotate_tac* −*6*)
  apply (*erule_tac* *x*=*j* **in** *allE*)
  apply *simp*
apply (*subgoaL_tac* ∀ *t3* ≤ (*t* + *3* + *k* + *d*). *a t3* = [])
  prefer *2*

apply (*simp add: ServiceCenter_def*)
apply *clarify*
apply (*case_tac t3*)
apply *simp+*
apply (*erule_tac x=Suc (t + 3 + k + d) in allE*)
apply *simp*
apply (*subgoal_tac*
   ∀ x ≤ d + d. ack (t + 3 + k + x) = [sending_data])
   prefer 2
   apply (*erule Gateway_L8, assumption+*)
   apply *clarify*
   apply (*rotate_tac 11*)
   apply (*erule_tac x=3 + k + j in allE*)
   apply *simp*
   apply (*subgoal_tac t + (3 + k + j) = t + 3 + k + j*)
      prefer 2
      apply *arith*
   apply *simp*
   apply *assumption+*
apply (*simp add: GatewaySystem_L1*)
done

B.7.5. Proof of the Refinement for the Gateway System

lemma *GatewaySystem_L0*:
GatewaySystem req dt stop lose d ack vc
⟹
GatewaySystemReq req dt stop lose d ack vc
apply (*simp add: GatewaySystemReq_def*)
apply (*simp add: GatewaySystem_def*)
apply *clarify*
apply (*subgoal_tac msg (Suc 0) a*)
   prefer 2
   apply (*simp add: ServiceCenter_a_msg*)
apply (*subgoal_tac*
GatewayReq req dt a stop lose d ack i vc)
   prefer 2
   apply (*simp add: Gateway_L0*)
apply (*case_tac dt (Suc t) = []*)
apply (*case_tac dt (Suc (Suc t)) = []*)
apply *simp*
apply *clarify*
apply (*simp add: GatewaySystem_L2*)
apply *simp*
apply *clarify*
apply (*simp add: GatewaySystem_L3*)
apply *simp*
apply (*case_tac dt (Suc (Suc t)) = []*)
apply *simp*
apply *clarify*
apply (*simp add: GatewaySystem_L3*)
apply *simp*
apply *clarify*
apply (*simp add: GatewaySystem_L3*)
done

B.7.6. Proof of the Refinement Relation for the Extended Gateway Requirements

lemma *Gateway_L0ext:*
Gateway req dt a stop lose d ack i vc
$\Longrightarrow$
GatewayReqExt req dt a stop lose d ack i vc
  apply (*simp add: GatewayReqExt_def*)
  apply *auto*
  apply (*simp add: Gateway_L1*)
  apply (*subgoal_tac*
    ∀ *t2 < t + 3 + k + d. i t2 = []*)
   prefer *2*
  apply (*simp add: Gateway_L7*)
  apply *simp*
  apply (*subgoal_tac*
    ∀ *y ≤ k. ack (t + y) = [connection_ok]*)
   prefer *2*
  apply (*simp add: Gateway_L6*)
  apply *simp*
  apply (*simp add: Gateway_L2*)
  apply (*simp add: Gateway_L3*)
  apply (*subgoal_tac*
    ∀ *x ≤ d + d. ack (t + x) = [sending_data]*)
   prefer *2*
  apply (*simp add: Gateway_L8*)
  apply *simp*
  apply (*simp add: Gateway_L4*)
  done

B.7.7. Lemma about Extended Gateway Requirements

lemma *GatewayReq_L1ext:*
⟦ *ServiceCenter i a; msg (Suc 0) req;*
  msg (Suc 0) stop; ts lose; msg (Suc 0) a;
  GatewayReqExt req dt a stop lose d ack i vc;
  inf_last_ti dt (t + 2 + k) ≠ [];
  ack t = [init_state]; req (Suc t) = [init];
  ∀ *t1≤t. req t1 = []; req (Suc (Suc t)) = [];*
  ∀ *m< k + 3. req (t + m) ≠ [send];*
  req (t + 3 + k) = [send];
  ∀ *j≤2 * d + (4 + k). lose (t + j) = [False]*⟧
$\Longrightarrow$ *vc (2 * d + (t + (4 + k))) = [vc_com]*

  apply (*subgoal_tac ack (t+2) = [connection_ok]*)
   prefer *2*
  apply (*simp add: GatewayReqExt_def*)
  apply (*erule_tac x=t* **in** *allE*)
  apply *clarify*
  apply (*rotate_tac −2*)
  apply (*erule_tac x=k* **in** *allE*)
  apply *simp*
  apply (*subgoal_tac*
    ∀ *j≤2 * d + (4 + k). lose (t + j) = [False]*)
   prefer *2*
   apply *simp*
  apply (*rotate_tac −1*)
  apply (*erule_tac x=Suc 0* **in** *allE*)

    apply (*rotate_tac* −3)
    apply (*erule_tac* x=*Suc* (*Suc 0*) **in** *allE*)
    apply *simp*

  apply (*subgoal_tac* ∀ *t2* < (t + *3* + k + d). *i t2* = [])
    prefer *2*
    apply (*simp add*: *GatewayReqExt_def*)
    apply (*erule_tac* x=t **in** *allE*)
    apply *clarify*
    apply (*rotate_tac* −2)
    apply (*erule_tac* x=k **in** *allE*)
    apply *simp*

  apply (*subgoal_tac ack* (t + *2* + k) = [*connection_ok*])
    prefer *2*
    apply (*simp add*: *GatewayReqExt_def*)
    apply (*erule_tac* x=t+*2* **in** *allE*)
    apply *clarify*
    apply (*rotate_tac* −1)
    apply (*erule_tac* x=k **in** *allE*)
    apply *simp*
    apply *clarify*
    apply (*subgoal_tac* ∀ m≤k. *req* (*Suc* (*Suc* (t + m))) ≠ [*send*])
      prefer *2*
      apply *clarify*
      apply (*rotate_tac* *10*)
      apply (*erule_tac* x=m+*2* **in** *allE*)
      apply *simp*
      apply (*subgoal_tac*
      ∀ j≤k. *lose* (*Suc* (*Suc* (t + j))) = [*False*])
        prefer *2*
        apply *clarify*
        apply (*rotate_tac* *12*)
        apply (*erule_tac* x=j+*2* **in** *allE*)
        apply *simp*
    apply *simp*

  apply (*subgoal_tac* ∀ j≤ d + *1*. *lose* (t + *3* + k + d + j) = [*False*])
    prefer *2*
    apply *clarify*
    apply (*rotate_tac* −5)
    apply (*erule_tac* x=*3* + k + d + j **in** *allE*)
    apply *simp*
    apply (*subgoal_tac*
    (t + (*3* + k + d + j)) = (t + *3* + k + d + j))
      prefer *2*
      apply *arith*
    apply *simp*

  apply (*subgoal_tac ack* ((t+*3*+k+d) + d) = [*sending_data*])
    prefer *2*
    apply (*subgoal_tac*
    ∀ x ≤ (d+d). *ack* (t + *3* + k + x) = [*sending_data*])
      prefer *2*
      apply (*subgoal_tac*
      ∀ j≤ d + d. *lose* (t + *3* + k + j) = [*False*])
        prefer *2*

    apply *clarify*
    apply (*rotate_tac −6*)
    apply (*erule_tac x=3 + k + j* **in** *allE*)
    apply *simp*
    apply (*subgoal_tac*
    t + (3 + k + j) = t + 3 + k + j)
      prefer *2*
      apply *simp*
    apply *simp*
  apply (*subgoal_tac ack (t + 3 + k) = [sending_data]*)
    prefer *2*
    apply (*subgoal_tac*
    ∀ *j≤ d + 1. lose (t + 2 + k + j) = [False]*)
      prefer *2*
      apply *clarify*
      apply (*rotate_tac −7*)
      apply (*erule_tac x=2 + k + j* **in** *allE*)
      apply *simp*
      apply (*subgoal_tac*
      Suc (Suc (t + (k + j))) = Suc (Suc (t + k + j)))
        prefer *2*
        apply *simp*
      apply *simp*
    apply (*simp add: GatewayReqExt_def*)
    apply (*erule_tac x=t+2+k* **in** *allE*)
    apply *clarify*
    apply (*rotate_tac −1*)
    apply (*erule_tac x=k* **in** *allE*)
    apply *clarify*
    apply (*simp add: nat_number*)
  apply (*subgoal_tac* ∀ *t3 ≤ t+3+k+d. a t3 = []*)
    prefer *2*
    apply (*simp add: ServiceCenter_def*)
    apply *clarify*
    apply (*rotate_tac −2*)
    apply (*erule_tac x=t3−(1::nat)* **in** *allE*)
    apply (*rotate_tac −7*)
    apply (*erule_tac x=(t3 − (1::nat))* **in** *allE*)
    apply (*case_tac t3 = 0*)
    apply *simp*
    apply (*subgoal_tac*
    (t3 − (1::nat)) < t + 3 + k + d)
      prefer *2*
      apply *arith*
    apply *simp*
  apply (*simp add: GatewayReqExt_def*)
  apply (*rotate_tac −1*)
  apply (*erule_tac x=d+d* **in** *allE*)
  apply *simp*

apply (*subgoal_tac a (t+4+k+d) = [sc_ack]*)
  prefer *2*
  apply (*simp add: GatewayReqExt_def*)
  apply (*erule_tac x=t+2+k* **in** *allE*)
  apply *clarify*
  apply (*rotate_tac −1*)
  apply (*erule_tac x=k* **in** *allE*)

apply *clarify*
apply (*subgoaltac*
$\forall ka \leq Suc\ d.\ lose\ (Suc\ (Suc\ (t + k + ka))) = [False]$)
  prefer *2*
  apply *clarify*
  apply (*rotate_tac 12*)
  apply (*erule_tac x=k + ka + 2 in allE*)
  apply (*subgoaltac*
    $Suc\ (Suc\ (t + (k + ka))) = Suc\ (Suc\ (t + k + ka))$)
    prefer *2*
    apply *simp*
  apply *simp*
apply (*subgoaltac i (Suc (t + 2 + k + d)) \neq []*)
  prefer *2*
  apply (*simp add: nat_number*)
apply (*erule_tac*
$V = ack\ (t + 2 + k) = [connection\_ok] \land req\ (Suc\ (t + 2 + k)) = [send]$
  $\land\ (\forall ka \leq Suc\ d.\ lose\ (t + 2 + k + ka) = [False]) \longrightarrow$
$i\ (Suc\ (t + 2 + k + d)) = inf\_last\_ti\ dt\ (t + 2 + k)$
$\land\ ack\ (Suc\ (t + 2 + k)) = [sending\_data]\ \textbf{in}\ thin\_rl$)
apply (*simp only: ServiceCenter_def*)
apply (*erule_tac x=t + 3 + k + d in allE*)
apply *clarify*
apply (*simp add: nat_number*)

apply (*simp add: GatewayReqExt_def*)
apply (*erule_tac x=t+3+k+d in allE*)
apply *clarify*
apply (*rotate_tac −1*)
apply (*erule_tac x=k in allE*)
apply *clarify*
apply *simp*
apply (*subgoaltac t + 4 + k + d = 4 + (t + (k + d))*)
  prefer *2*
  apply *simp*
apply (*subgoaltac*
$4 + (2 * d + (t + k)) = 2 * d + (t + (4 + k))$)
  prefer *2*
  apply *simp*
apply *simp*
done

B.7.8. Proof of the Refinement for the Gateway System (Based on the Extended Gateway Requirements)

lemma *GatewaySystem_L0ext*:
GatewaySystem req dt stop lose d ack vc
$\Longrightarrow$
GatewaySystemReq req dt stop lose d ack vc
apply (*simp add: GatewaySystemReq_def*)
apply (*simp add: GatewaySystem_def*)
apply *clarify*
apply (*subgoaltac msg (Suc 0) a*)
  prefer *2*
  apply (*simp add: ServiceCenter_a_msg*)
apply (*subgoaltac*
GatewayReqExt req dt a stop lose d ack i vc)

prefer *2*
apply (*simp add: Gateway_L0ext*)
apply (*case_tac dt (Suc t) = []*)
apply (*case_tac dt (Suc (Suc t)) = []*)
apply *simp*
apply *clarify*
apply (*subgoal_tac*
inf_last_ti dt (t + 2 + k) ≠ [])
  prefer *2*
  apply (*simp add: inf_last_ti_nonempty_k*)
apply (*simp add: GatewayReq_L1ext*)
apply *simp*
apply *clarify*
apply (*subgoal_tac inf_last_ti dt (t + 2 + k) ≠ []*)
  prefer *2*
  apply (*subgoal_tac dt (t+2) ≠ []*)
    prefer *2*
    apply *simp*
  apply (*erule inf_last_ti_nonempty*)
apply (*simp add: GatewayReq_L1ext*)
apply *simp*
apply (*case_tac dt (Suc (Suc t)) = []*)
apply *simp*
apply *clarify*
apply (*subgoal_tac inf_last_ti dt (t + 1 + k) ≠ []*)
  prefer *2*
  apply (*subgoal_tac dt (t+1) ≠ []*)
    prefer *2*
    apply *simp*
  apply (*erule inf_last_ti_nonempty*)
apply (*subgoal_tac*
inf_last_ti dt (t + 2 + k) ≠ [])
  prefer *2*
  apply (*simp add: inf_last_ti_nonempty_k*)
apply (*simp add: GatewayReq_L1ext*)
apply *simp*
apply *clarify*
apply (*subgoal_tac inf_last_ti dt (t + 2 + k) ≠ []*)
  prefer *2*
  apply (*subgoal_tac dt (t+2) ≠ []*)
    prefer *2*
    apply *simp*
  apply (*erule inf_last_ti_nonempty*)
apply (*simp add: GatewayReq_L1ext*)
done

end